VOLUME ONE HUNDRED A

ADVANCES IN GENETICS

unknown. Originally, the importance of virulence genes was assessed by testing the ability of *Agrobacterium* insertion mutants for different genes to induce tumor formation after infection of plants highly susceptible to *Agrobacterium* (usually tobacco or kalanchoe) (Horsch et al., 1986; Stachel, An, Flores, & Nester, 1985). It is likely that non-essential or accessory genes play a role in the infection of other, less susceptible, plant species. In addition, the importance of these genes may be undetectable under laboratory conditions, but play a role under natural conditions, in the highly competitive environment of the rhizosphere. Besides the *vir* genes present on the Ti-plasmid, some chromosomally encoded genes are required for *Agrobacterium* virulence. These genes are involved mostly in sensing environmental conditions and modulation of virulence induction (*chv*G, *chv*E, *chv*I, *exo*R), or in the attachment of *Agrobacterium* cells to plant cell/tissue surface (*chv*A, *chv*B, *exo*C).

3. Virulence factors in non-*Agrobacterium* species

Interestingly, homologs of virulence genes have been found in many bacterial species related to *Agrobacterium*, within plasmids of bacteria belonging to the *Rhizobiaceae* family. Indeed, the study of many plasmids from species of the *agrobacteria-rhizobia* complex shows that partial or complete *vir* regions are often found in these plasmids and that combination of these mobile genetic elements in one strain may lead to the assembling of a functional DNA transfer machinery (Weisberg et al., 2020), which raises the question of whether other, non-*Agrobacterium*, species can transfer DNA to eukaryotic hosts. Because most of the virulence determinants are found on a mobile genetic element (Ti-plasmid), a strain harboring no Ti-plasmid or only an incomplete virulence system may become virulent by acquiring a plasmid from another strain by conjugative plasmid transfer.

It has been known for a long time that transferring a plasmid(s) containing a functional virulence region and a T-DNA to several species closely related to *Agrobacterium* (belonging to the *Rhizobiaceae* and *Phyllobacteriaceae* families) could confer the ability to transfer T-DNA to the recipient species (Broothaerts et al., 2005; Hooykaas, Klapwijk, Nuti, Schilperoort, & Rorsch, 1977; Wendt, Doohan, & Mullins, 2012; Zuniga-Soto, Mullins, & Dedicova, 2015). These results indicate that these bacterial strains harbor all the chromosomally encoded factors required for T-DNA transfer. More recently, it was shown that a *vir* region of the native plasmid p42a of *Rhizobium etli* CFN42 strain is functional, and able to transfer T-DNA to host

VOLUME ONE HUNDRED AND TEN

ADVANCES IN GENETICS

ELSEVIER

ACADEMIC PRESS

An imprint of Elsevier

Academic Press is an imprint of Elsevier
50 Hampshire Street, 5th Floor, Cambridge, MA 02139, United States
525 B Street, Suite 1650, San Diego, CA 92101, United States
The Boulevard, Langford Lane, Kidlington, Oxford OX5 1GB, United Kingdom
125 London Wall, London, EC2Y 5AS, United Kingdom

First edition 2022

ISBN: 978-0-443-18878-7
ISSN: 0065-2660

For information on all Academic Press publications
visit our website at https://www.elsevier.com/books-and-journals

Publisher: Zoe Kruze
Developmental Editor: Cindy Angelita Gardose
Production Project Manager: Abdulla Sait
Cover Designer: Greg Harris

Typeset by STRAIVE, India

Contents

Contributors

Vitaly Citovsky
Department of Biochemistry and Cell Biology, State University of New York, Stony Brook, NY, United States

Vanessa Desantis
Department of Biomedical Sciences and Human Oncology, Pharmacology Section, University of Bari Aldo Moro Medical School, Bari, Italy

Benoit Lacroix
Department of Biochemistry and Cell Biology, State University of New York, Stony Brook, NY, United States

Domenico Ribatti
Department of Basic Medical Sciences, Neurosciences and Sensory Organs, University of Bari Medical School, Bari, Italy

Antonio G. Solimando
Department of Biomedical Sciences and Human Oncology (DIMO), Unit of Internal Medicine and Clinical Oncology, University of Bari Aldo Moro Medical School, Bari, Italy

CHAPTER ONE

Genetic factors governing bacterial virulence and host plant susceptibility during *Agrobacterium* infection

Benoit Lacroix* and Vitaly Citovsky
Department of Biochemistry and Cell Biology, State University of New York, Stony Brook, NY, United States
*Corresponding author: e-mail address: benoit.lacroix@stonybrook.edu

Contents

Abstract

Several species of the *Agrobacterium* genus represent unique bacterial pathogens able to genetically transform plants, by transferring and integrating a segment of their own DNA (T-DNA, transferred DNA) in their host genome. Whereas in nature this process results in uncontrolled growth of the infected plant cells (tumors), this capability of *Agrobacterium* has been widely used as a crucial tool to generate transgenic plants,

Advances in Genetics, Volume 110
ISSN 0065-2660
https://doi.org/10.1016/bs.adgen.2022.08.001

for research and biotechnology. The virulence of *Agrobacterium* relies on a series of virulence genes, mostly encoded on a large plasmid (Ti-plasmid, tumor inducing plasmid), involved in the different steps of the DNA transfer to the host cell genome: activation of bacterial virulence, synthesis and export of the T-DNA and its associated proteins, intracellular trafficking of the T-DNA and effector proteins in the host cell, and integration of the T-DNA in the host genomic DNA. Multiple interactions between these bacterial encoded proteins and host factors occur during the infection process, which determine the outcome of the infection. Here, we review our current knowledge of the mechanisms by which bacterial and plant factors control *Agrobacterium* virulence and host plant susceptibility.

1. Introduction

Agrobacterium tumefaciens was discovered more than a century ago as the causing agent of the crown-gall disease, which results in uncontrolled cell division (tumors) mostly at the roots and base of the stem of the infected plants (Kado, 2014). Different species or strains of *Agrobacterium* may cause different diseases in various host plants: *A. tumefaciens* causes crown galls, *A. vitis* causes galls on the stem of grape species and *A. rhizogenes* causes root proliferation (hairy root). The infection of plants by *Agrobacterium* species represents a unique case of active horizontal gene transfer in the living world. Indeed, *Agrobacterium* virulence relies mostly on two essential regions of its Ti-plasmid (tumor-inducing plasmid): the virulence region (containing the *vir* genes) and the T-DNA region (containing the T-DNA, a segment of DNA transferred and integrated into the genome of the host plant cell). In wild-type *Agrobacterium* strains, the T-DNA contains several genes, which will be expressed in the transformed plant cells. Some of these gene products will affect host cell division and cause ectopic growth (such as the crown gall), while another series of genes encode proteins responsible for the synthesis of small molecules (opines) that are exported out of the host tissues and used by *Agrobacterium* as a source of nutrition (Escobar & Dandekar, 2003). Because the T-DNA transfer does not depend on its sequence, T-DNA genes can be replaced by any sequence of interest, which made possible the use of *Agrobacterium* as a tool for plant genetic transformation. Since the discovery of the T-DNA as the "tumor-inducing agent," numerous studies have focused on identifying the bacterial and plant genetic factors involved in the infection process and on deciphering the molecular mechanism of plant genetic transformation mediated by *Agrobacterium* (Gelvin, 2003a; Lacroix & Citovsky, 2019). Most of our current knowledge of the infection mechanisms is derived from the experiments performed with *A. tumefaciens*; thus, we will use the term *Agrobacterium* in this chapter, although we will refer mostly to *A. tumefaciens*.

Susceptibility of plants to *Agrobacterium* infection varies widely between plant species; most notably monocotyledon species are generally recalcitrant to transformation (De Cleene & De Ley, 1976). Variations are also observed within a species, between varieties, or accessions (Chateau, Sangwan, & Sangwan-Norreel, 2000). Moreover, organs, tissues, or cell types of a plant differ in their susceptibility, and different treatments (nutritional, hormonal) of the plant cells or tissues affect transformation efficiency, suggesting that the physiological status of the cells also alters their susceptibility. Under laboratory conditions, *Agrobacterium*-mediated genetic transformation can be achieved with most plant species, with variable efficiency, and non-plant species (yeast, fungi, and even mammalian cells) can also be transformed (Lacroix, Tzfira, Vainstein, & Citovsky, 2006). Many genetic factors affecting the susceptibility of plants to *Agrobacterium* have been identified (Gelvin, 2003b; Lacroix & Citovsky, 2019); their presence and activity may represent the determinants for the outcome of the infection process. In this review, we will focus on our current understanding of bacterial and plant factors controlling *Agrobacterium* virulence and host plant susceptibility, from the first cellular interactions to the integration of T-DNA into the host genome.

2. Bacterial factors defining *Agrobacterium* virulence

Agrobacterium virulence relies mostly on a series of genes (the *vir* genes), located on the Ti plasmid, of which expression may be activated in response to plant and environmental signals. These genes encode proteins involved at different stages of plant infection and can be classified into different groups according to their level of requirement for *Agrobacterium*-mediated transformation (Table 1). The first group represents a core of essential genes, absolutely required for infection: *vir*A and *vir*G (main system for *vir* gene induction), the *vir*B operon, and *vir*D4 (export of macromolecules from the bacterial cells), and *vir*D1 and *vir*D2 (synthesis and transport of the T-DNA). The second group of *vir* genes can be defined as important but not absolutely essential (transformation occurs only at a very low rate with *Agrobacterium* mutated in these genes): *vir*C1 and *vir*C2 (T-DNA synthesis), *vir*E1 and *vir*E2 (protection and nuclear import of the T-DNA). A third group, sometimes qualified as host-range genes, corresponds to genes that

Table 1 *Agrobacterium* virulence genes (see text for references).

Requirement for transformation	Gene name	Function
Essential	*vir*A	Sensor, part of the two-component regulator of *vir* gene expression
	*vir*B1–11	Main components of the T4SS responsible for export of T-complex and effector proteins
	*vir*D1	T-strand synthesis
	*vir*D2	T-strand synthesis, protein component of the T-complex, nuclear import of the T-DNA
	*vir*D4	Part of the T4SS (coupling factor)
	*vir*G	Part of two-component system, transcription activator of *vir* genes
Important	*vir*C1–2	Enhance T-strand synthesis
	*vir*E1	Chaperone for VirE2 protein
	*vir*E2	Effector protein, protection of the T-strand
Host range	*vir*D3	Unknown
	*vir*D5	Effector protein, prevents VirF degradation
	*vir*E3	Effector protein, interacts with VirE2, anchor for VirE2 after its entry in the host cell, transcription regulator
	*vir*F	Effector protein, F-box protein, proteasomal degradation of several host plant target proteins
Unknown	*vir*H1–2	Detoxification
	*vir*J	Homolog of chromosomally encoded *acv*B
	*vir*K	Unknown
	*vir*L	Unknown
	*vir*M	Unknown

may be required as an enhancer of transformation efficiency only with a certain host or in specific conditions; this group includes *vir*D5, *vir*E3, and *vir*F. Finally, some of the *vir* genes are not found in all *Agrobacterium* strains (*vir*D3, *vir*H1 and *vir*H 2, *vir*J, *vir*K, *vir*L, *vir*M); although they belong to the *vir* regulon, their potential role in *Agrobacterium* infection is generally plant cells (albeit with low efficiency), when that bacterial strain was transformed with a plasmid containing a T-DNA region only, (Lacroix & Citovsky, 2016). It was later shown that *R. etli vir* gene expression was regulated by phenolics, similarly to *Agrobacterium* (Wang, Lacroix, Guo, & Citovsky, 2017).

4. Regulation of virulence gene expression

All the *vir* gene or operon promoters contain at least one specific sequence, 10- to 12-bp sequences (*vir* box) located between 200 and 50bp upstream of the transcription initiation site, which is required for the coordinated *vir* gene induction (Steck, Morel, & Kado, 1988). The activity of these promoters, and thus the expression of the *vir* genes, is mostly under the control of a two-component receptor system composed of the VirA and VirG proteins (Stachel & Zambryski, 1986). VirA is an integral membrane sensor protein that integrates several signals; once activated, VirA mediates the phosphorylation of VirG, which then binds to the *vir* box containing promoters and promotes *vir* gene expression (Fig. 1). The

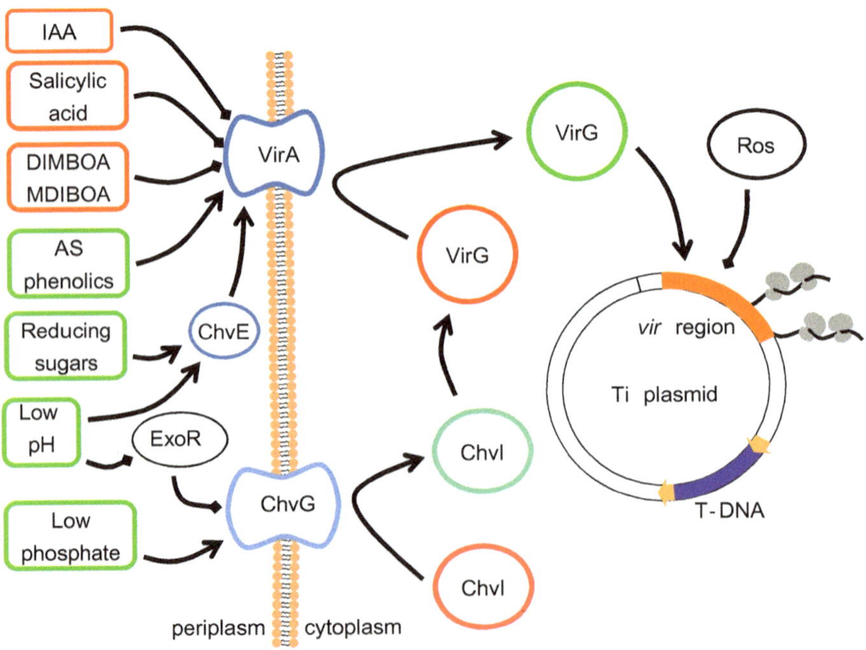

Fig. 1 Regulation of *vir* gene expression by plant and environmental factors (see text for details).

most important of these activating signals is a plant-produced phenolic compound, acetosyringone (AS, 3,5-dimethoxyacetophenone) (Bolton, Nester, & Gordon, 1986; Stachel, Messens, Van Montagu, & Zambryski, 1985). Several other phenolic compounds related to AS (including phenolic glycosides) are also able to activate the VirA/VirG system (Joubert et al., 2004; Melchers, Regensburg-Tuink, Schilperoort, & Hooykaas, 1989). Genetic studies suggest that AS and other phenolics bind directly to VirA (Lee, Jin, Sim, & Nester, 1995), although it cannot be completely excluded that a yet unknown intermediary protein bind phenolics in the periplasm before activating VirA. A range of reducing monosaccharides (e.g., D-glucose and D-galactose) can bind to ChvE, a chromosome-encoded periplasmic protein, which then enhances *vir* gene expression by directly binding to VirA (Cangelosi, Ankenbauer, & Nester, 1990; Shimoda, Toyoda-Yamamoto, Aoki, & Machida, 1993). This interaction results in an increase of VirA/VirG sensitivity and of saturating concentration for the *vir* gene induction by phenolics (Shimoda et al., 1990). Other environmental features, such as low pH and low phosphate concentration, also affect *vir* gene expression. Together, they activate ChvG/ChvI, another two-component regulatory system, which in turn increases the expression of *virG* (Charles & Nester, 1993). Low pH results in the degradation of ExoR, a periplasmic inhibitor of ChvG (Heckel, Tomlinson, Morton, Choi, & Fuqua, 2014). Moreover, low pH (between pH 5 and 6) enhances VirA activity directly (Melchers et al., 1989) or through ChvE (Gao & Lynn, 2005).

Because the expression of *vir* genes is costly in energy for the bacterial cells (Platt, Bever, & Fuqua, 2012), it is also important that their expression is repressed in later infectious stages when Vir proteins are not required anymore. Several mechanisms could play this role in *Agrobacterium*. Indeed, *Agrobacterium* virulence is inhibited in response to the auxin IAA (indole acetic acid), produced at high levels by developing *Agrobacterium*-induced tumors (Liu & Nester, 2006). IAA likely can bind VirA, acting as a competitive inhibitor of AS. Another pathway capable of turning off the *vir* gene expression was recently suggested. In this pathway, sucrose would bind to and inactivate *Agrobacterium* SghR, resulting in the expression of SghA; SghA would then free SA (salicylic acid) from its storage form SAG (SA β-glucoside), and SA would inhibit VirA (Wang et al., 2019). However, this model relies on two yet unproved assumptions: the massive release of sucrose from plants to the extracellular space and the export of SAG (usually stored in vacuoles) from plant cells.

In addition, the ability to inhibit *vir* gene expression may represent a mechanism of plant defense to prevent infection by *Agrobacterium*, and consequently a source of variability of susceptibility to *Agrobacterium* between plant species. For example, two chemicals isolated from corn seedling homogenates (DIMBOA and MDIBOA) were shown to inhibit both *vir* gene induction and *Agrobacterium* growth (Sahi, Chilton, & Chilton, 1990; Zhang et al., 2000). Like IAA, these two molecules derive from the tryptophan biosynthetic pathway (Melanson, Chilton, Masters-Moore, & Chilton, 1997), and could also inhibit *vir* gene expression by binding to VirA. In studies with plants deficient or overexpressing genes for the synthesis of salicylic acid (SA, the major signal molecule of the systemic acquired resistance pathway), it was shown that SA inhibited *vir* gene expression, most likely by interfering with VirA activity (Anand et al., 2008; Yuan et al., 2007). The plant gaseous growth regulator ethylene was also able to inhibit *Agrobacterium* virulence (Nonaka, Sugawara, Minamisawa, Yuhashi, & Ezura, 2008; Nonaka, Yuhashi, et al., 2008), although it is not known if ethylene has a direct effect on *vir* gene expression.

Whereas most of the regulation pathways cited above converge on the VirA/VirG two-component system, *vir* gene expression can also be altered via pathways independent of this system. Indeed, a mutation in Ros, a transcription regulator encoded by the *Agrobacterium* chromosome, resulted in the activation of expression of the *vir*C and *vir*D operons (Close et al., 1987). Furthermore, an investigation of the small RNA-dependent gene regulations in *Agrobacterium* showed that some of the *vir* genes were regulated via this pathway (Dequivre et al., 2015).

5. T-DNA synthesis

The T-DNA synthesis corresponds to the production of the single strand T-DNA segment via a mechanism of strand-replacement (Stachel, Timmerman, & Zambryski, 1986). Two 24–25 bp sequences, known as the left border (LB) and the right border (RB), present as a direct repeat on the Ti-plasmid, mark the beginning and the end of the T-DNA (Peralta & Ream, 1985; Yadav, Vanderleyden, Bennett, Barnes, & Chilton, 1982), which can be mobilized from the Ti-plasmid in the form of a single-stranded DNA intermediate (the T-strand). The processing of the T-strand is mediated by VirD2, acting as an endonuclease (Albright, Yanofsky, Leroux, Ma, & Nester, 1987; Yanofsky et al., 1986), associated with VirD1, most likely acting as a DNA topoisomerase (Ghai & Das, 1989). The result of this process is the immature T-complex, comprised of the T-strand and of a molecule of the VirD2 protein that remains covalently linked to its 5′-end (RB) (Young & Nester, 1988). VirC1 and VirC2 were also shown to bind to "overdrive" sequences located close to the T-DNA borders, causing an increase in the amount of processed T-strand molecules (De Vos & Zambryski, 1989).

6. T-DNA and associated protein export

Export of T-DNA and associated proteins from the bacterial cells relies on a Type IV secretion system (T4SS) encoded by the *vir*B operon and *vir*D4, via a mechanism close to the plasmid translocation during bacterial conjugation (Li & Christie, 2018). In addition to the VirD2-T-strand complex, four other proteins are translocated to the host cell: VirE2, VirE3, VirD5, and VirF. Firstly, the VirD2-T-strand complex and the translocated proteins must be targeted to the T4SS within the bacterial cell. All these proteins display an arginine-rich C-terminal sequence required for their export (Vergunst et al., 2000, 2005), although it is not an exact signal sequence (identical for all these proteins). The presence of a similar signal sequence shared between the different exported proteins suggests a unique pathway of recognition of the translocated proteins within bacterial cells, but such a unique pathway has not been identified yet. Instead, several different factors have been suggested for each of the exported proteins. Interaction with the T4SS coupling protein VirD4 seemed required to recruit VirE2 to the cell poles, where T4SS is assembled (Atmakuri, Ding, & Christie, 2003). VirC1 and VirC2 likely assist the targeting of VirD2 (and thus the T-complex) to the cell poles (Atmakuri, Cascales, Burton, Banta, & Christie, 2007). More recently, VBPs (VirD2-binding proteins) were identified as bacterial factors able to recruit VirD2 and the associated T-strand to the T4SS energizing components (VirD4, VirB4, and VirB11) (Guo, Hou, Hew, & Pan, 2007; Guo, Jin, Sun, Hew, & Pan, 2007). In other systems, VBPs can recruit relaxase proteins and conjugating DNA intermediates to the T4SS during conjugation. However, VBPs did not interact with the other exported effector proteins of *Agrobacterium*, thus they do not represent the only factors recognizing C-terminal sequences of the exported proteins.

Agrobacterium represents a model for the study of T4SSs; thus, the structure of its T4SS is known in detail (Christie, 2004), and the sequence of interactions between the different subunits of the T4SS and the T-DNA transport substrate was elucidated (Cascales & Christie, 2004).

7. Attachment and biofilm formation

Generally, interactions between pathogenic bacteria and their eukaryotic host require a close interaction between bacteria and the plant cells/tissue is necessary. The attachment of the bacterial cell relies on several bacterial and plant factors and often results in the formation of a biofilm. In the case of *Agrobacterium*, this process of transition from motile planktonic bacterial cells to biofilm can be divided into three steps (Heindl et al., 2014). First, bacterial cells are attracted toward the plant cell surface by chemotaxis, which is triggered by the same plant exudates that induce virulence, i.e., phenolic compounds and reducing sugars (Guo, Huang, & Yang, 2017). Bacteria approach the surface of plant tissue via flagellum-dependent motility, relying on the ChvA sensor (Merritt, Danhorn, & Fuqua, 2007; Wright, Deakin, & Shaw, 1998). Second, there is initial contact and reversible attachment between the bacterial cells and the host tissue surface. Third, bacterial attachment is stabilized, and bacteria are embedded within a biofilm. *Agrobacterium* synthesizes several exocellular polysaccharides known to play a role in attachment and biofilm formation: 1,2-β-D-glucan produced and exported by the activity of ChvA, ChvB, and ExoC, for attachment and virulence (Cangelosi et al., 1989; de Iannino & Ugalde, 1989); unipolar polysaccharides (UPPs) involved in attachment (Xu, Kim, Danhorn, Merritt, & Fuqua, 2012); and cellulose for attachment consolidation and biofilm formation (Matthysse, 1983). Potential plant cell surface receptors able to bind these exopolysacharides (such as the lectins known to be involved in *Rhizobium*-host cell attachment) have not yet been identified for *Agrobacterium* host cell attachment. Proteins exposed at the surface of *Agrobacterium* cells could also play a role in attachment. Indeed, several *Arabidopsis* genes were shown to encode proteins interacting with VirB2, the main component of the T4SS pilus (Hwang & Gelvin, 2004). Although T-DNA transfer efficiency was affected in *Arabidopsis* lines mutants in these genes, it is not known whether this change in efficiency reflects a disruption of attachment or of another step (passage of T-DNA or proteins from bacterial to plant cell, or cell signaling). VirB5 is a minor component of the T4SS located at the tip of the VirB2 pilus (Aly & Baron, 2007); although it was shown that the addition of free extracellular VirB5 enhances the T-DNA transfer (Lacroix & Citovsky, 2011), there is no indication that it is involved in attachment. Whereas their requirement for virulence is not clear, CtpA and PilA (two other extracellular *Agrobacterium* proteins) were shown to play a role in the first steps of *Agrobacterium* attachment to plant tissue surface (Wang, Haitjema, & Fuqua, 2014).

Two Arabidopsis mutant lines resistant to *Agrobacterium* could potentially be deficient in proteins involved in bacterial attachment, although their function was not completely elucidated. This is the case of CSLA9 (Zhu, Nam, Carpita, Matthysse, & Gelvin, 2003), encoding an enzyme that is likely synthesizing cell wall polysaccharides, and of AGP17 (an extracellular arabinogalactan-protein) (Gaspar et al., 2004).

8. T-DNA entry in the plant cell

The mechanism of the entry of T-DNA and associated proteins through the host cell membrane and the nature of the plant factors involved in this process are not completely understood. Theoretically, there are three ways by which the exported macromolecules could pass through the membrane and enter the host cell cytoplasm. Similar to a mechanism occurring in the type III secretion system (T3SS) (Notti & Stebbins, 2016), the transported macromolecules could pass through the VirB2 pilus, their entry into the plant cell relying on interactions between pilus proteins and host-cell membrane-associated proteins. In another scenario, close to a proposed mechanism for bacterial conjugation (Cabezón, Ripoll-Rozada, Peña, de la Cruz, & Arechaga, 2015), after depolymerization of the VirB2 pilus, the membranes of the bacterial and host cells would be close enough to fusion together, allowing the transfer of macromolecules. Finally, the macromolecules exported from *Agrobacterium* cells could be deposited at the surface of the host cell and internalized via interactions with host membrane factors and potentially via the endocytosis pathway. Recent studies have shown that the transport of VirE2 probably takes advantage of the host cell endocytosis pathway (Li & Pan, 2017); indeed, VirE2 entry into plant cells was associated with early endosome formation, and VirE2 interacted with AP2M, a protein located on the cytoplasmic side of the clathrin-coated vesicles. So far, there is no evidence that the internalization of other translocated macromolecules is also associated with the endocytotic pathway.

9. T-DNA intracellular transport

The transport of the T-DNA within the plant cell relies on interactions between translocated Vir proteins and several host factors. Prior to the potential integration of the T-DNA in the host genome, the

VirD2-T-strand complex and translocated proteins playing a role in the host nucleus must be imported into the nucleus. The T-complex depends on its associated proteins to reach its host cell nucleus. VirD2, attached covalently to the 5′-end of the T-strand, is targeted to the nucleus via its interaction with a specific importin alpha (Ballas & Citovsky, 1997), and other members of the importin alpha family (Bakó, Umeda, Tiburcio, Schell, & Koncz, 2003). VirD2 likely represents the "pilot" protein mediating the targeting of T-complex toward the nucleus. However, the transport of a large DNA molecule (such as the T-DNA) through the cytoplasm and the nuclear pore likely requires more than a single protein molecule. Many studies suggest that VirE2, as an ssDNA-binding protein, can bind and coat the T-DNA, efficiently leading to the formation of a mature T-complex (Citovsky, Wong, & Zambryski, 1989; Gelvin, 1998). Indeed, VirE2 binds to ssDNA with a strong affinity (Christie, Ward, Winans, & Nester, 1988; Citovsky et al., 1989), which results in ssDNA-VirE2 filament with a helical structure (Abu-Arish et al., 2004). Transformation experiments using *Agrobacterium vir*E2 mutant strains result in an increased level of truncations in the integrated T-DNA (Rossi, Hohn, & Tinland, 1996), consistent with the role of VirE2 in protecting the T-DNA against nucleolytic degradation. However, so far, such a mature T-complex has not been visualized in living plant cells, suggesting that it might not exist as a stable macromolecular assembly. Early studies showed a nuclear targeting for VirE2 labeled with different markers (Citovsky, Zupan, Warnick, & Zambryski, 1992; Ziemienowicz, Görlich, Lanka, Hohn, & Rossi, 1999). In later studies, VirE2 fused with autofluorescent proteins remained mostly cytoplasmic (Lee, Fang, Kuang, & Gelvin, 2008; Shi, Lee, & Gelvin, 2014), with a tendency to form aggregates consistent with the strong VirE2 homopolymerization. In a different experimental setting, where VirE2 fusion with a partial GFP sequence was expressed in *Agrobacterium* and fluorescence was reconstituted after *Agrobacterium*-mediated transfer in plant cell expressing the other part of GFP, VirE2 was at least partially nuclear (Li, Yang, Tu, Lim, & Pan, 2014). The intracellular targeting of VirE2 may depend on its interaction with several host proteins. Indeed, VirE2 was shown to interact with VIP1 (VirE2 interacting protein 1) (Tzfira, Vaidya, & Citovsky, 2001), VIP2 (VirE2 interacting protein 2) (Anand et al., 2007), core histones (Lacroix, Loyter, & Citovsky, 2008; Loyter et al., 2005), and importins alpha (Bhattacharjee et al., 2008). Other translocated proteins (VirE3, VirD5, and VirF) are also targeted to the nucleus, after binding to host importins. Interestingly, VirE3 interacts with VirE2, which is likely involved in two steps of VirE2 intracellular transport:

accumulation of VirE2 on the cytoplasmic side of the host cell plasma membrane immediately after it enters the host cell (Li, Tu, & Pan, 2018), and VirE2 subsequent nuclear targeting (Lacroix, Vaidya, Tzfira, & Citovsky, 2005). A more recent study suggests that VirE2 nuclear import depends on the presence of the T-DNA and that VirE2 interaction with the plant nucleoporin CG1 facilitates the passage of the T-complex through nuclear pores (Li et al., 2020).

10. T-DNA integration

Several studies have shown that the sites of integration of *Agrobacterium* T-DNA are not targeted to a specific region of the host genome. In the first studies, T-DNA integration sites were found throughout the genome but mostly in the region of the active expression (Alonso et al., 2003). However, this observation was based on an experimental bias because the transgenic plants were recovered after selection and required expression of the gene encoding resistance agent to the selection pressure. Indeed, a similar experiment realized without selection pressure revealed that integration occurs truly randomly in the host genome (Kim, Veena, & Gelvin, 2007), whereas there is a possible local bias toward some epigenetic markers in the chromatin (Shilo et al., 2017). *Agrobacterium* does not encode a dedicated integrase among its effector proteins, thus the integration of the T-DNA into the host genome depends on the activity of several host pathways. Indeed, VirD2 was once suspected to mediate T-DNA integration (Pansegrau, Schoumacher, Hohn, & Lanka, 1993; Tinland, Schoumacher, Gloeckler, Bravo-Angel, & Hohn, 1995), but it was later shown not to be the case (Ziemienowicz, Tinland, Bryant, Gloeckler, & Hohn, 2000). It is likely, however, that interactions between *Agrobacterium* effector and host proteins are required for integration. It has been shown that induction of DSBs in plant tissue prior to *Agrobacterium* infection results in an increase in stable transformation (Salomon & Puchta, 1998), suggesting that DSBs in the host genome might be a target for T-DNA integration and that the activation of DNA damage reaction following DSB induction might also enhance T-DNA integration. Experiments realized with yeast (*Saccharomyces cerevisiae*) as host cells demonstrate that the outcome of integration indeed depends on the activity of host factors. Using yeast as host cells has two advantages: many viable mutants in different DSB repair pathways are available, and DSB repair occurs via either HR (homologous recombination) or NHEJ (non-homologous end joining) pathways. With yeast mutants in genes essential for the HR pathway (Rad51 or Rad52), only integration via NHEJ was observed, whereas when the NHEJ was disrupted (mutation in Ku70 or Mre11) all integration events resulted from HR (van Attikum, Bundock, & Hooykaas, 2001; van Attikum et al., 2003). In plants (where DSB repair is mostly mediated by NHEJ), the situation seems more complex. Conflicted data resulted from the analysis of T-DNA integration efficiency in *Arabidopsis* lines mutated in genes representing the different DSB repair pathways. In *Arabidopsis* mutant deficient in Ku80 and Lig4, it was found that T-DNA integration was inhibited in two studies (Friesner & Britt, 2003; Li et al., 2005), while it was reported in another study that mutation in Ku80 did not affect integration efficiency (Gallego, Bleuyard, Daoudal-Cotterell, Jallut, & White, 2003). In a similar study with *Arabidopsis* mutants in different genes representing the known pathway of DSB repair, it was reported that none of these mutants displayed inhibition of the T-DNA integration (Park et al., 2015). However, only very low levels of integration were observed in *Arabidopsis* lines disrupted in several of these pathways (Mestiri, Norre, Gallego, & White, 2014). In rice plants where expression of Ku70, Ku80, and Lig4 was downregulated, lower rates of T-DNA integration were measured (Nishizawa-Yokoi et al., 2012). Redundancy between the different DNA repair pathways existing in plants may explain partially the discrepancy between the results reported in these different studies. It is also possible that other pathways are involved in T-DNA integration.

Recently, the role of polymerase theta (also known as Tebichi in plants) in T-DNA integration was investigated. Indeed, sequence analysis of a large number of T-DNA integration sites showed that many of them displayed a signature of ligation by polymerase theta, and T-DNA integration was impaired in an *Arabidopsis* line deficient in this gene (van Kregten et al., 2016). Because polymerase theta, first discovered as a suppressor of genome instability, is also involved in microhomology-mediated end joining (MMEJ), or alternative end-joining (alt-EJ) (Black, Kashkina, Kent, & Pomerantz, 2016), it was suspected that this alternative pathway of DNA repair is involved in T-DNA integration. In a more recent study, it was shown that in addition to polymerase theta ligating the 3′-end of the T-DNA to the genomic DNA, other proteins were required to remove the VirD2 protein from the T-complex before the ligation (Kralemann et al., 2022). Indeed, while attachment of the 3′-end resulted from the polymerase theta activity exclusively, removal of VirD2 and ligation of the 5′-end could occur via two different mechanisms. Either VirD2 was removed by TDP2 and ligation was mediated by canonical NHEJ, or VirD2 removal relied on MRE11 (part of the MRN complex), and attachment to the genomic DNA occurred via the polymerase theta pathway.

Moreover, in *Arabidopsis* double mutants in TDP2 and Mre11 very low rate of stable transformation was observed, suggesting the requirement for several DNA repair pathways for T-DNA integration.

11. Host factors interacting with *Agrobacterium* effector proteins

Besides the main roles of *Agrobacterium* essential effectors for intracellular transport and, potentially, integration, many other bacterial-host protein interactions occur during the infection process (Table 2). These interactions may represent either the targets of *Agrobacterium* effector proteins (their interaction being part of a mechanism by which the effector

Table 2 *Agrobacterium* encoded effector proteins and their host interacting proteins (see text for references).

Effector	Host protein	Host species	Known or suspected functions
VirD2	Importin α	Arabidopsis	Nuclear import
	Cyclophilins	Arabidopsis	Unknown
	2C protein phosphatase	Tomato	Regulation of nuclear import
	CAK2M	Alfalfa	Interaction with host chromatin, integration
	TBP	Alfalfa	
	Core histones	Yeast	Interaction with host chromatin
VirE2	Importin α	Arabidopsis	Nuclear import
	VIP1	Arabidopsis	Nuclear import, plant transcriptional regulation
	RSG (VIP1 ortholog)	Tobacco	
	bZIP proteins (related to VIP1)	Arabidopsis	
	VIP2	Arabidopsis benthamiana	Required for integration, plant transcriptional regulation
	XRCC4	Arabidopsis	Integration
	GST	Rice	Regulation of nuclear import

Continued

Table 2 *Agrobacterium* encoded effector proteins and their host interacting proteins (see text for references).—cont'd

Effector	Host protein	Host species	Known or suspected functions
VirE3	Importin α	Arabidopsis	Nuclear import
	Csn5	Arabidopsis	Plant transcriptional regulation
	Brp	Arabidopsis	
	JAZ8	Arabidopsis	Regulation of plant defense reaction
VirD5	Spt4	Yeast	Mitotic destabilization
	Aurora kinase	Yeast and Arabidopsis	
	VIP1	Arabidopsis	Prevents VBF binding and VIP1 proteasomal degradation
	VIP2	Arabidopsis	Plant transcriptional regulation
VirF	ASK1	Arabidopsis	Proteasomal degradation (part of the SCF complex)
	VIP1	Arabidopsis	Target for degradation, T-complex uncoating, plant transcriptional regulation
	VFP3	Arabidopsis	Plant transcriptional regulation
	VFP5	Arabidopsis	Plant transcriptional regulation
	VFP4	Arabidopsis	Target for degradation, plant transcriptional regulation

protein enhances the efficiency of *Agrobacterium*-mediated transformation) or pathways of plant "defense" against *Agrobacterium* (resulting in inhibition of *Agrobacterium*-mediated transformation). Until now, the search for host factors involved in *Agrobacterium*-plant interactions was done mostly with susceptible plant species (such as *Arabidopsis* and *Nicotiana* species). It is likely that if more investigations were performed with resistant host plants, more of the second category host factors would be uncovered.

11.1 VirD2

VirD2 interacts with several members of a subgroup of *Arabidopsis* cyclophilins, single domain cyclophilins (Bakó et al., 2003; Deng et al., 1998). Cyclophilins represent a large family of proteins (21 in *Arabidopsis*), present in different subcellular compartments and involved in a wide variety of cellular processes (Romano, Horton, & Gray, 2004). Their potential function in *Agrobacterium* T-DNA transfer is not known, whereas treatment with cyclosporine A (which binds cyclophilins and inhibited interaction between cyclophilin and VirD2) resulted in inhibition of T-DNA transfer. A type 2C serine/threonine protein phosphatase from tomato was also found to interact with VirD2 (Tao, Rao, Bhattacharjee, & Gelvin, 2004), this phosphatase is thought to inhibit VirD2 nuclear targeting via dephosphorylation of VirD2. Furthermore, it was reported that VirD2 interacts and is phosphorylated by CAK2M (a cyclin-dependent kinase-activating kinase) in alfalfa cells; CAK2M also phosphorylates a subunit of RNA-polymerase II able to recruit TATA-binding protein (TBP), and VirD2 was found to associate with TBP. Because orthologs of CAK2M and TBP are involved in transcription-coupled DNA repair, it was suggested that they are involved in T-DNA integration. Finally, VirD2 interacted with yeast core histones (Wolterink-van Loo, Escamilla Ayala, Hooykaas, & van Heusden, 2015), which might also play a role in the interaction between T-complex and host chromatin during integration.

11.2 VirE2

The first series of studies uncovered the interaction between VirE2 and VIP1 (VirE2 interacting protein 1) by yeast-two-hybrid screening (Kunik et al., 2001); and that increased transformation levels were observed in tobacco plants overexpressing *Arabidopsis* VIP1 (AtVIP1), likely by the role of VIP1 in facilitating VirE2 nuclear targeting (Tzfira et al., 2001; Tzfira, Vaidya, & Citovsky, 2002). However, a later study, using *Arabidopsis* mutants, concluded that VIP1 was not required for the *Agrobacterium*-mediated transformation (Shi et al., 2014). VIP1 belongs to a family of bZIP (basic Leucine zipper) transcription factors found in *Arabidopsis* and most plant species. Recently, it was shown that VirE2 could interact not only with AtVIP1 but also with several of its close *Arabidopsis* homologs, as well as with the AtVIP1 tobacco ortholog NtRSG (Wang et al., 2018). Moreover, VirE2 proteins from different *Agrobacterium* strains displayed variable binding efficiency with the different AtVIP1 homologs. Another study confirmed the interactions between VirE2 and several AtVIP1 homologs and demonstrated that disrupting VIP1 transcription activator ability (but not its VirE2 binding ability) did not affect T-DNA transfer efficiency (Lapham et al., 2018). The second interactor of VirE2 (VIP2, or VirE2 interacting protein 2) was discovered after yeast two-hybrid screening of an *Arabidopsis* cDNA library. It was demonstrated that VIP2 (a transcription regulator) is required for *Agrobacterium* T-DNA integration (stable transformation) but not for its transient expression, both in *Arabidopsis* and *N. benthamiana* (Anand et al., 2007). VirE2 also interacted with *Arabidopsis* XRCC4 (X-ray cross complementation group 4), a protein involved in the NHEJ DNA repair pathway, potentially interfering with the host plant DSB repair pathway to facilitate the T-DNA integration (Vaghchhipawala, Vasudevan, Lee, Morsy, & Mysore, 2012). Finally, it was reported that VirE2 interacted with the rice protein OsGSTU5, a tau class GST (glutathione S-transferase). That interaction occurred in the host cell cytoplasm and likely resulted in VirE2 glutathionylation and inhibition of *Agrobacterium*-mediated transformation, potentially by reducing VirE2 affinity for single-stranded T-DNA (Tiwari et al., 2022).

11.3 VirD5

Agrobacterium VirD5 was first found to interact with VirF, another effector protein; VirF being naturally unstable in the plant cell, its association with VirD5 resulted in the protection of VirF against degradation (Magori & Citovsky, 2011). Moreover, VirD5 expression was shown to induce cell toxicity in yeast and plant cells via its interaction with kinetochore proteins (Spt4 in yeast and Aurora kinases in yeast and plants), causing chromosomal instability (Zhang & Hooykaas, 2019; Zhang, van Heusden, & Hooykaas, 2017). How this interaction could be involved in the *Agrobacterium* infection process is still unknown. It was also reported that VirD5 interacted with VIP1 (VirE2 interacting protein 1), competing with VBF and potentially preventing VIP1 degradation (Wang et al., 2014), as well as with VIP2 (VirE2 interacting protein 2), potentially preventing VIP2 interaction with Cap-binding proteins (proteins involved in mRNA biosynthesis) (Wang et al., 2018). In the latter case, it is not clear how this interaction would play a role in *Agrobacterium* T-DNA transfer and integration.

11.4 VirE3

Besides binding to VirE2 and being targeted to the host cell nucleus via the interaction with the importin alpha (García-Rodríguez, Schrammeijer, & Hooykaas, 2006; Lacroix et al., 2005), VirE3 was shown to interact with several Arabidopsis proteins, such as Csn5 (a component of the COP9 signalosome) and Brp (a member of the TFIIB family that binds to TATA box) (García-Rodríguez et al., 2006). Although a potential role of CSn5 in *Agrobacterium*-mediated T-DNA transfer is not known, the interaction with Brp correlates with the demonstrated function of VirE3 in the transcriptional regulation (Niu, Zhou, Henkel, van Heusden, & Hooykaas, 2015). Recently, it was reported that VirE3 interacts with *Arabidopsis* JAZ8, a transcription regulator involved in the plant defense reaction (Li et al., 2021). It seems that VirE3 was able to interfere with JAZ8 transcription regulator activity to mitigate plant defense response via the SA pathway, whereas overexpression of JAZ8 reduced the efficiency of *Agrobacterium*-mediated transformation.

11.5 VirF

VirF contains an F-box domain and was shown to interact with several ASK proteins (plant homologs of Skp1 functioning in the SCF pathway of proteasomal degradation) from *Arabidopsis* (Schrammeijer et al., 2001). VirF activity as a component of the SCF pathway was demonstrated in yeast and plant cells, and VirF interacted with one of its target *Arabidopsis* VIP1 (Tzfira, Vaidya, & Citovsky, 2004). Indeed, VirF was able to induce destabilization of VIP1 (and of its associated VirE2, potentially stripping the T-DNA from VirE2 coating) via proteasomal degradation. Later, several other targets of VirF were discovered, including VFP4 a transcription regulator involved in the plant defense response (García-Cano, Hak, Magori, Lazarowitz, & Citovsky, 2018); VirF also interacted with two closely related trihelix-domain transcription factors (VFP3 and VFP5) but without activating the host UPS pathway (García-Cano et al., 2015). Interestingly, *Agrobacterium* induces the expression of a plant F-box protein (VBF, VIP1 binding F-box), which could partially substitute for VirF activity in plant cells by targeting VIP1 (Zaltsman, Krichevsky, Loyter, & Citovsky, 2010). It was originally believed that, unlike the *Agrobacterium* octopine strain A6, the nopaline strain C58 did not encode a functional VirF. However, it was shown that C58-VirF is most likely a functional F-box protein because it contains an F-box domain and was able to bind ASK1 (Lacroix & Citovsky, 2015). C58-VirF did not interact with VIP1, suggesting that it could have a set of target host proteins different from A6-VirF.

12. Plant transcriptional response to *Agrobacterium* infection

Transcriptomic studies have shown that the expression of many host genes is modified upon *Agrobacterium* infection. Among the genes of which expression is regulated, many of them are involved in plant defense reactions (reviewed in (Willig, Duan, & Zhang, 2018)).

Like most of the higher eukaryotes, plants can sense bacteria through the perception of PAMPs (pathogen-associated molecular patterns), via a receptor that triggers a plant defense response when activated. For example, *Arabidopsis* was able to detect *Agrobacterium* via one of these PAMPs called EF-Tu (elongation factor thermo unstable), resulting in a defense response (Zipfel et al., 2006). Plants usually recognize another PAMP, flagellin 22 (flg22, a fragment of bacterial flagellin) via the receptor FLS2 (flagellin sensitive 2), but *Agrobacterium* harbors a highly divergent flg22 that evades detection by most plants (Felix, Duran, Volko, & Boller, 1999). Interestingly, some plant species, such as the wild grape *Vitis riparia*, may encode a different FLS2 receptor that can recognize *Agrobacterium* flg22; tobacco plants expressing this receptor displayed increased resistance to *Agrobacterium* (Fürst et al., 2020). Several elements of the signaling cascade involved in pathogen-triggered immunity in *Arabidopsis* were shown to be important for *Agrobacterium* infection. Indeed, two mitogen-activated protein kinase kinases MKK4/MKK5 and their downstream mitogen-activated protein kinases MPK3/MPK6 were shown to play a crucial role in the induction of various plant defense pathways by *Agrobacterium*, and their activity affected the efficiency of *Agrobacterium*-mediated transformation (Liu et al., 2021).

Although a plant defense response to *Agrobacterium* infection is observed at the transcriptional level, in most plant species no extensive defense reaction is observed upon *Agrobacterium* infection. As shown in the previous section, several of the translocated effectors were suggested to act as transcriptional regulators themselves or to interfere with host transcription regulation pathways, which could mitigate the plant defense reaction. For example, VirE2 interaction with two plant transcription regulators involved in plant defense reaction (VIP1 and VIP2) might alter their activity. VirE3 was shown to act as a transcription regulator itself, as well as to interact with Arabidopsis JAZ8 (affecting JAZ8 role as a SA pathway activator). VirD5 is also suspected to induce a transcriptional response, either directly or via its interaction with VIP2. Finally, VirF can interact with several transcription regulators, and in some cases (e.g., VIP1, VFP4) induce their proteasomal degradation. In fact, it was demonstrated that VirF-dependent degradation of VFP4 resulted in mitigating the host defense response. Globally, *Agrobacterium* encodes effector proteins that can in many ways interfere with the transcriptional response of its host plant.

For the most part, the plant transcriptional response to *Agrobacterium* seems to be related to the regulation of the general plant defense response.

In a few cases, however, the activation of a plant factor specifically involved in *Agrobacterium*-mediated T-DNA transfer was shown. For example, VirE3 induced the expression of *Arabidopsis* VBF, a functional equivalent of *Agrobacterium* VirF (Niu et al., 2015). It was reported in two recent studies that several genes involved in plant DNA repair pathways are transcriptionally activated in tobacco and *Arabidopsis* plants challenged by *Agrobacterium* infection, and this activation seemed to depend on the presence of the *vir* genes (Hu, Lacroix, & Citovsky, 2021; Joseph, Chandhini, Das, Mysore, & Shah, 2021). We can speculate that this increased expression of DNA repair-related genes plays a role in T-DNA integration.

13. Conclusions

The study of *Agrobacterium*-mediated genetic transformation of plants has revealed a complex network of interactions between bacterial and host factors. On the bacterial side, *Agrobacterium* encoded proteins are either directly mediating the transfer of T-DNA or enhancing this transfer, for example, by interfering with plant defense response. On the plant side, many plant proteins interact with bacterial factors, and a plant defense reaction is triggered. *Agrobacterium* has evolved many strategies to use host pathways to its advantage and to mitigate the plant defense response. Most research so far was performed in plant species susceptible to *Agrobacterium*, and it is likely that more host factors will be discovered in the future, particularly those that render some plant species resistant to *Agrobacterium*. Furthermore, there is a certain level of variability of the virulence factors between *Agrobacterium* strains and species, to which correspond different capabilities of the factors from different host species to interact with bacterial factors. The adequation and the nature of these interactions between host and bacterial species determine the outcome of infection and are responsible for the differences in susceptibility of host plants to different *Agrobacterium* strains.

Acknowledgments

The work in the VC laboratory is supported by grants from USDA/NIFA, NIH, NSF, and BARD to V.C.

Disclosure statement

The authors are not aware of any affiliations, membership, funding, or financial holdings that might be perceived as affecting the objectivity of this review.

References

Abu-Arish, A., Frenkiel-Krispin, D., Fricke, T., Tzfira, T., Citovsky, V., Wolf, S. G., et al. (2004). Three-dimensional reconstruction of *Agrobacterium* VirE2 protein with single-stranded DNA. *The Journal of Biological Chemistry, 279*, 25359–25363.

Albright, L. M., Yanofsky, M. F., Leroux, B., Ma, D. Q., & Nester, E. W. (1987). Processing of the T-DNA of *Agrobacterium tumefaciens* generates border nicks and linear, single-stranded T-DNA. *Journal of Bacteriology, 169*, 1046–1055.

Alonso, J. M., Stepanova, A. N., Leisse, T. J., Kim, C. J., Chen, H., Shinn, P., et al. (2003). Genome-wide insertional mutagenesis of *Arabidopsis thaliana*. *Science, 301*, 653–657.

Aly, K. A., & Baron, C. (2007). The VirB5 protein localizes to the T-pilus tips in *Agrobacterium tumefaciens*. *Microbiology, 153*, 3766–3775.

Anand, A., Krichevsky, A., Schornack, S., Lahaye, T., Tzfira, T., Tang, Y., et al. (2007). *Arabidopsis* VIRE2 INTERACTING PROTEIN2 is required for *Agrobacterium* T-DNA integration in plants. *Plant Cell, 19*, 1695–1708.

Anand, A., Uppalapati, S. R., Ryu, C. M., Allen, S. N., Kang, L., Tang, Y., et al. (2008). Salicylic acid and systemic acquired resistance play a role in attenuating crown gall disease caused by *Agrobacterium tumefaciens*. *Plant Physiology, 146*, 703–715.

Atmakuri, K., Cascales, E., Burton, O. T., Banta, L. M., & Christie, P. J. (2007). *Agrobacterium* ParA/MinD-like VirC1 spatially coordinates early conjugative DNA transfer reactions. *The EMBO Journal, 26*, 2540–2551.

Atmakuri, K., Ding, Z., & Christie, P. J. (2003). VirE2, a type IV secretion substrate, interacts with the VirD4 transfer protein at cell poles of *Agrobacterium tumefaciens*. *Molecular Microbiology, 49*, 1699–1713.

Bakó, L., Umeda, M., Tiburcio, A. F., Schell, J., & Koncz, C. (2003). The VirD2 pilot protein of *Agrobacterium*-transferred DNA interacts with the TATA box-binding protein and a nuclear protein kinase in plants. *Proceedings of the National Academy of Sciences of the United States of America, 100*, 10108–10113.

Ballas, N., & Citovsky, V. (1997). Nuclear localization signal binding protein from *Arabidopsis* mediates nuclear import of *Agrobacterium* VirD2 protein. *Proceedings of the National Academy of Sciences of the United States of America, 94*, 10723–10728.

Bhattacharjee, S., Lee, L. Y., Oltmanns, H., Cao, H., Veena, C., & J., and Gelvin, S. B. (2008). IMPa-4, an *Arabidopsis* importin alpha isoform, is preferentially involved in *Agrobacterium*-mediated plant transformation. *Plant Cell, 20*, 2661–2680.

Black, S. J., Kashkina, E., Kent, T., & Pomerantz, R. T. (2016). DNA polymerase theta: A unique multifunctional end-joining machine. *Genes*, 7, 67.

Bolton, G. W., Nester, E. W., & Gordon, M. P. (1986). Plant phenolic compounds induce expression of the *Agrobacterium tumefaciens* loci needed for virulence. *Science, 232*, 983–985.

Broothaerts, W., Mitchell, H. J., Weir, B., Kaines, S., Smith, L. M., Yang, W., et al. (2005). Gene transfer to plants by diverse species of bacteria. *Nature, 433*, 629–633.

Cabezón, E., Ripoll-Rozada, J., Peña, A., de la Cruz, F., & Arechaga, I. (2015). Towards an integrated model of bacterial conjugation. *FEMS Microbiology Reviews, 39*, 81–95.

Cangelosi, G. A., Ankenbauer, R. G., & Nester, E. W. (1990). Sugars induce the *Agrobacterium* virulence genes through a periplasmic binding protein and a transmembrane signal protein. *Proceedings of the National Academy of Sciences of the United States of America, 87*, 6708–6712.

Cangelosi, G. A., Martinetti, G., Leigh, J. A., Lee, C. C., Thienes, C., & Nester, E. W. (1989). Role for *Agrobacterium tumefaciens* ChvA protein in export of beta-1,2-glucan. *Journal of Bacteriology, 171*, 1609–1615.

Cascales, E., & Christie, P. J. (2004). Definition of a bacterial type IV secretion pathway for a DNA substrate. *Science, 304*, 1170–1173.

Charles, T. C., & Nester, E. W. (1993). A chromosomally encoded two-component sensory transduction system is required for virulence of *Agrobacterium tumefaciens*. *Journal of Bacteriology*, *175*, 6614–6625.

Chateau, S., Sangwan, R. S., & Sangwan-Norreel, B. S. (2000). Competence of *Arabidopsis thaliana* genotypes and mutants for *Agrobacterium tumefaciens*-mediated gene transfer: Role of phytohormones. *Journal of Experimental Botany*, *51*, 1961–1968.

Christie, P. J. (2004). Type IV secretion: The *Agrobacterium* VirB/D4 and related conjugation systems. *Biochimica et Biophysica Acta*, *1694*, 219–234.

Christie, P. J., Ward, J. E., Winans, S. C., & Nester, E. W. (1988). The *Agrobacterium tumefaciens virE2* gene product is a single-stranded-DNA-binding protein that associates with T-DNA. *Journal of Bacteriology*, *170*, 2659–2667.

Citovsky, V., Wong, M. L., & Zambryski, P. C. (1989). Cooperative interaction of *Agrobacterium* VirE2 protein with single stranded DNA: Implications for the T-DNA transfer process. *Proceedings of the National Academy of Sciences of the United States of America*, *86*, 1193–1197.

Citovsky, V., Zupan, J., Warnick, D., & Zambryski, P. C. (1992). Nuclear localization of *Agrobacterium* VirE2 protein in plant cells. *Science*, *256*, 1802–1805.

Close, T. J., Rogowsky, P. M., Kado, C. I., Winans, S. C., Yanofsky, M. F., & Nester, E. W. (1987). Dual control of *Agrobacterium tumefaciens* Ti plasmid virulence genes. *Journal of Bacteriology*, *169*, 5113–5118.

De Cleene, M., & De Ley, J. (1976). The host range of crown gall. *The Botanical Review*, *42*, 389–466.

de Iannino, N. I., & Ugalde, R. A. (1989). Biochemical characterization of avirulent *Agrobacterium tumefaciens chvA* mutants: Synthesis and excretion of beta-(1-2)glucan. *Journal of Bacteriology*, *171*, 2842–2849.

De Vos, G., & Zambryski, P. C. (1989). Expression of *Agrobacterium* nopaline specific VirD1, VirD2, and VirC1 proteins and their requirement for T-strand production in *E. coli*. *Molecular Plant-Microbe Interactions*, *2*, 43–52.

Deng, W., Chen, L., Wood, D. W., Metcalfe, T., Liang, X., Gordon, M. P., et al. (1998). *Agrobacterium* VirD2 protein interacts with plant host cyclophilins. *Proceedings of the National Academy of Sciences of the United States of America*, *95*, 7040–7045.

Dequivre, M., Diel, B., Villard, C., Sismeiro, O., Durot, M., Coppée, J. Y., et al. (2015). Small RNA deep-sequencing analyses reveal a new regulator of virulence in *Agrobacterium fabrum* C58. *Molecular Plant-Microbe Interactions*, *28*, 580–589.

Escobar, M. A., & Dandekar, A. M. (2003). Agrobacterium tumefaciens as an agent of disease. *Trends in Plant Science*, *8*, 380–386.

Felix, G., Duran, J. D., Volko, S., & Boller, T. (1999). Plants have a sensitive perception system for the most conserved domain of bacterial flagellin. *The Plant Journal*, *18*, 265–276.

Friesner, J., & Britt, A. B. (2003). Ku80- and DNA ligase IV-deficient plants are sensitive to ionizing radiation and defective in T-DNA integration. *The Plant Journal*, *34*, 427–440.

Fürst, U., Zeng, Y., Albert, M., Witte, A. K., Fliegmann, J., & Felix, G. (2020). Perception of *Agrobacterium tumefaciens* flagellin by FLS2(XL) confers resistance to crown gall disease. *Nature Plants*, *6*, 22–27.

Gallego, M. E., Bleuyard, J. Y., Daoudal-Cotterell, S., Jallut, N., & White, C. I. (2003). Ku80 plays a role in non-homologous recombination but is not required for T-DNA integration in *Arabidopsis*. *The Plant Journal*, *35*, 557–565.

Gao, R., & Lynn, D. G. (2005). Environmental pH sensing: Resolving the VirA/VirG two-component system inputs for *Agrobacterium* pathogenesis. *Journal of Bacteriology*, *187*, 2182–2189.

García-Cano, E., Hak, H., Magori, S., Lazarowitz, S. G., & Citovsky, V. (2018). The *Agrobacterium* F-box protein effector VirF destabilizes the *Arabidopsis* GLABROUS1 enhancer/binding protein-like transcription factor VFP4, a transcriptional activator of defense response genes. *Molecular Plant-Microbe Interactions, 31*, 576–586.

García-Cano, E., Magori, S., Sun, Q., Zhang, S., Lazarowitz, S. G., & Citovsky, V. (2015). Interaction of *Arabidopsis* trihelix-domain transcription factors VFP3 and VFP5 with *Agrobacterium* virulence protein VirF. *PLoS One, 10*, e014212.

García-Rodríguez, F. M., Schrammeijer, B., & Hooykaas, P. J. J. (2006). The *Agrobacterium* VirE3 effector protein: A potential plant transcriptional activator. *Nucleic Acids Research, 34*, 6496–6504.

Gaspar, Y. M., Nam, J., Schultz, C. J., Lee, L. Y., Gilson, P. R., Gelvin, S. B., et al. (2004). Characterization of the *Arabidopsis* lysine-rich arabinogalactan-protein *AtAGP17* mutant (*rat1*) that results in a decreased efficiency of *Agrobacterium* transformation. *Plant Physiology, 135*, 2162–2171.

Gelvin, S. B. (1998). *Agrobacterium* VirE2 proteins can form a complex with T strands in the plant cytoplasm. *Journal of Bacteriology, 180*, 4300–4302.

Gelvin, S. B. (2003a). *Agrobacterium*-mediated plant transformation: The biology behind the "gene-jockeying" tool. *Microbiology and Molecular Biology Reviews, 67*, 16–37.

Gelvin, S. B. (2003b). Improving plant genetic engineering by manipulating the host. *Trends in Biotechnology, 21*, 95–98.

Ghai, J., & Das, A. (1989). The *virD* operon of *Agrobacterium tumefaciens* Ti plasmid encodes a DNA-relaxing enzyme. *Proceedings of the National Academy of Sciences of the United States of America, 86*, 3109–3113.

Guo, M., Hou, Q., Hew, C. L., & Pan, S. Q. (2007). *Agrobacterium* VirD2-binding protein is involved in tumorigenesis and redundantly encoded in conjugative transfer gene clusters. *Molecular Plant-Microbe Interactions, 20*, 1201–1212.

Guo, M., Huang, Z., & Yang, J. (2017). Is there any crosstalk between the chemotaxis and virulence induction signaling in *Agrobacterium tumefaciens*? *Biotechnology Advances, 35*, 505–511.

Guo, M., Jin, S., Sun, D., Hew, C. L., & Pan, S. Q. (2007). Recruitment of conjugative DNA transfer substrate to *Agrobacterium* type IV secretion apparatus. *Proceedings of the National Academy of Sciences of the United States of America, 104*, 20019–20024.

Heckel, B. C., Tomlinson, A. D., Morton, E. R., Choi, J. H., & Fuqua, C. (2014). *Agrobacterium tumefaciens exoR* controls acid response genes and impacts exopolysaccharide synthesis, horizontal gene transfer, and virulence gene expression. *Journal of Bacteriology, 196*, 3221–3233.

Heindl, J. E., Wang, Y., Heckel, B. C., Mohari, B., Feirer, N., & Fuqua, C. (2014). Mechanisms and regulation of surface interactions and biofilm formation in *Agrobacterium*. *Frontiers in Plant Science, 5*, 176.

Hooykaas, P. J. J., Klapwijk, P. M., Nuti, M. P., Schilperoort, R. A., & Rorsch, A. (1977). Transfer of the *Agrobacterium tumefaciens* Ti plasmid to avirulent agrobacteria and to *Rhizobium ex planta*. *Journal of General Microbiology, 98*, 477–484.

Horsch, R. B., Klee, H. J., Stachel, S., Winans, S. C., Nester, E. W., Rogers, S. G., et al. (1986). Analysis of *Agrobacterium tumefaciens* virulence mutants in leaf discs. *Proceedings of the National Academy of Sciences of the United States of America, 83*, 2571–2575.

Hu, Y., Lacroix, B., & Citovsky, V. (2021). Modulation of plant DNA damage response gene expression during *Agrobacterium* infection. *Biochemical and Biophysical Research Communications, 554*, 7–12.

Hwang, H. H., & Gelvin, S. B. (2004). Plant proteins that interact with VirB2, the *Agrobacterium tumefaciens* pilin protein, mediate plant transformation. *Plant Cell, 16*, 3148–3167.

Joseph, J. T., Chandhini, S., Das, S., Mysore, K. S., & Shah, J. M. (2021). Methylation status of *Arabidopsis* DNA repair gene promoters during *Agrobacterium* infection reveals epigenetic changes in three generations. *Plant Molecular Biology Reporter, 39*, 773–791.

Joubert, P., Beaupère, D., Wadouachi, A., Chateau, S., Sangwan, R. S., & Sangwan-Norreel, B. S. (2004). Effect of phenolic glycosides on *Agrobacterium tumefaciens virH* gene induction and plant transformation. *Journal of Natural Products, 67*, 348–351.

Kado, C. I. (2014). Historical account on gaining insights on the mechanism of crown gall tumorigenesis induced by *Agrobacterium tumefaciens*. *Frontiers in Microbiology, 5*, 340.

Kim, S. I., Veena, & Gelvin, S. B. (2007). Genome-wide analysis of *Agrobacterium* T-DNA integration sites in the *Arabidopsis* genome generated under non-selective conditions. *The Plant Journal, 51*, 779–791.

Kralemann, L. E. M., de Pater, S., Shen, H., Kloet, S. L., van Schendel, R., Hooykaas, P. J. J., et al. (2022). Distinct mechanisms for genomic attachment of the 5′ and 3′ ends of *Agrobacterium* T-DNA in plants. *Nature Plants, 8*, 526–534.

Kunik, T., Tzfira, T., Kapulnik, Y., Gafni, Y., Dingwall, C., & Citovsky, V. (2001). Genetic transformation of HeLa cells by *Agrobacterium*. *Proceedings of the National Academy of Sciences of the United States of America, 98*, 1871–1876.

Lacroix, B., & Citovsky, V. (2011). Extracellular VirB5 enhances T-DNA transfer from *Agrobacterium* to the host plant. *PLoS One, 6*, e25578.

Lacroix, B., & Citovsky, V. (2015). Nopaline-type Ti plasmid of *Agrobacterium* encodes a VirF-like functional F-box protein. *Scientific Reports, 5*, 16610.

Lacroix, B., & Citovsky, V. (2016). A functional bacterium-to-plant DNA transfer machinery of *Rhizobium etli*. *PLoS Pathogens, 12*, e1005502.

Lacroix, B., & Citovsky, V. (2019). Pathways of DNA transfer to plants from *Agrobacterium tumefaciens* and related bacterial species. *Annual Review of Phytopathology, 57*, 231–251.

Lacroix, B., Loyter, A., & Citovsky, V. (2008). Association of the *Agrobacterium* T-DNA-protein complex with plant nucleosomes. *Proceedings of the National Academy of Sciences of the United States of America, 105*, 15429–15434.

Lacroix, B., Tzfira, T., Vainstein, A., & Citovsky, V. (2006). A case of promiscuity: *Agrobacterium*'s endless hunt for new partners. *Trends in Genetics, 22*, 29–37.

Lacroix, B., Vaidya, M., Tzfira, T., & Citovsky, V. (2005). The VirE3 protein of *Agrobacterium* mimics a host cell function required for plant genetic transformation. *The EMBO Journal, 24*, 428–437.

Lapham, R. A., Lee, L. Y., Tsugama, D., Lee, S., Mengiste, T., & Gelvin, S. B. (2018). VIP1 and its homologs are not required for *Agrobacterium*-mediated transformation, but play a role in *Botrytis* and salt stress responses. *Frontiers in Plant Science, 9*, 749.

Lee, L. Y., Fang, M. J., Kuang, L. Y., & Gelvin, S. B. (2008). Vectors for multi-color bimolecular fluorescence complementation to investigate protein-protein interactions in living plant cells. *Plant Methods, 4*, 24.

Lee, Y. W., Jin, S., Sim, W. S., & Nester, E. W. (1995). Genetic evidence for direct sensing of phenolic compounds by the VirA protein of *Agrobacterium tumefaciens*. *Proceedings of the National Academy of Sciences of the United States of America, 92*, 12245–12249.

Li, Y. G., & Christie, P. J. (2018). The *Agrobacterium* VirB/VirD4 T4SS: Mechanism and architecture defined through *in vivo* mutagenesis and chimeric systems. *Current Topics in Microbiology and Immunology, 418*, 233–260.

Li, X., & Pan, S. Q. (2017). *Agrobacterium* delivers VirE2 protein into host cells *via* clathrin-mediated endocytosis. *Science Advances, 3*, e1601528.

Li, X., Tu, H., & Pan, S. Q. (2018). *Agrobacterium* delivers anchorage protein VirE3 for companion VirE2 to aggregate at host entry sites for T-DNA protection. *Cell Reports, 25*, 302–311.

Li, J., Vaidya, M., White, C., Vainstein, A., Citovsky, V., & Tzfira, T. (2005). Involvement of KU80 in T-DNA integration in plant cells. *Proceedings of the National Academy of Sciences of the United States of America, 102*, 19231–19236.

Li, S., Xu, B., Niu, X., Lu, X., Cheng, J., Zhou, M., et al. (2021). JAZ8 interacts with VirE3 attenuating *Agrobacterium*-mediated root tumorigenesis. *Frontiers in Plant Science, 12*, 685533.

Li, X., Yang, Q., Peng, L., Tu, H., Lee, L. Y., Gelvin, S. B., et al. (2020). *Agrobacterium*-delivered VirE2 interacts with host nucleoporin CG1 to facilitate the nuclear import of VirE2-coated T complex. *Proceedings of the National Academy of Sciences of the United States of America, 117*, 26389–26397.

Li, X., Yang, Q., Tu, H., Lim, Z., & Pan, S. Q. (2014). Direct visualization of *Agrobacterium*-delivered VirE2 in recipient cells. *The Plant Journal, 77*, 487–495.

Liu, T., Cao, L., Cheng, Y., Ji, J., Wei, Y., Wang, C., et al. (2021). MKK4/5-MPK3/6 cascade regulates *Agrobacterium*-mediated transformation by modulating plant immunity in *Arabidopsis*. *Frontiers in Plant Science, 12*, 731690.

Liu, P., & Nester, E. W. (2006). Indoleacetic acid, a product of transferred DNA, inhibits *vir* gene expression and growth of *Agrobacterium tumefaciens* C58. *Proceedings of the National Academy of Sciences of the United States of America, 103*, 4658–4662.

Loyter, A., Rosenbluh, J., Zakai, N., Li, J., Kozlovsky, S. V., Tzfira, T., et al. (2005). The plant VirE2 interacting protein 1. A molecular link between the *Agrobacterium* T-complex and the host cell chromatin? *Plant Physiology, 138*, 1318–1321.

Magori, S., & Citovsky, V. (2011). *Agrobacterium* counteracts host-induced degradation of its F-box protein effector. *Science Signaling, 4*. ra69.

Matthysse, A. G. (1983). Role of bacterial cellulose fibrils in *Agrobacterium tumefaciens* infection. *Journal of Bacteriology, 154*, 906–915.

Melanson, D., Chilton, M. D., Masters-Moore, D., & Chilton, W. S. (1997). A deletion in an indole synthase gene is responsible for the DIMBOA-deficient phenotype of *bxbx* maize. *Proceedings of the National Academy of Sciences of the United States of America, 94*, 13345–13350.

Melchers, L. S., Regensburg-Tuink, A. J., Schilperoort, R. A., & Hooykaas, P. J. J. (1989). Specificity of signal molecules in the activation of *Agrobacterium* virulence gene expression. *Molecular Microbiology, 3*, 969–977.

Merritt, P. M., Danhorn, T., & Fuqua, C. (2007). Motility and chemotaxis in *Agrobacterium tumefaciens* surface attachment and biofilm formation. *Journal of Bacteriology, 189*, 8005–8014.

Mestiri, I., Norre, F., Gallego, M. E., & White, C. I. (2014). Multiple host-cell recombination pathways act in *Agrobacterium* mediated transformation of plant cells. *The Plant Journal, 77*, 511–520.

Nishizawa-Yokoi, A., Nonaka, S., Saika, H., Kwon, Y. I., Osakabe, K., & Toki, S. (2012). Suppression of Ku70/80 or Lig4 leads to decreased stable transformation and enhanced homologous recombination in rice. *The New Phytologist, 196*, 1048–1059.

Niu, X., Zhou, M., Henkel, C. V., van Heusden, G. P., & Hooykaas, P. J. J. (2015). The *Agrobacterium tumefaciens* virulence protein VirE3 is a transcriptional activator of the F-box gene VBF. *The Plant Journal, 84*, 914–924.

Nonaka, S., Sugawara, M., Minamisawa, K., Yuhashi, K., & Ezura, H. (2008). 1-Aminocyclopropane-1-carboxylate deaminase enhances *Agrobacterium tumefaciens*-mediated gene transfer into plant cells. *Applied and Environmental Microbiology, 74*, 2526–2528.

Nonaka, S., Yuhashi, K., Takada, K., Sugaware, M., Minamisawa, K., & Ezura, H. (2008). Ethylene production in plants during transformation suppresses *vir* gene expression in *Agrobacterium tumefaciens*. *The New Phytologist, 178*, 647–656.

Notti, R. Q., & Stebbins, C. E. (2016). The structure and function of type III secretion systems. *Microbiology Spectrum, 4*. VMBF-0004-2015.

Pansegrau, W., Schoumacher, F., Hohn, B., & Lanka, E. (1993). Site-specific cleavage and joining of single-stranded DNA by VirD2 protein of *Agrobacterium tumefaciens* Ti plasmids: Analogy to bacterial conjugation. *Proceedings of the National Academy of Sciences of the United States of America, 90*, 11538–11542.

Park, S. Y., Vaghchhipawala, Z., Vasudevan, B., Lee, L. Y., Shen, Y., Singer, K., et al. (2015). *Agrobacterium* T-DNA integration into the plant genome can occur without the activity of key non-homologous end-joining proteins. *The Plant Journal, 81,* 934–946.

Peralta, E. G., & Ream, L. W. (1985). T-DNA border sequences required for crown gall tumorigenesis. *Proceedings of the National Academy of Sciences of the United States of America, 82,* 5112–5116.

Platt, T. G., Bever, J. D., & Fuqua, C. (2012). A cooperative virulence plasmid imposes a high fitness cost under conditions that induce pathogenesis. *Proceedings of the Royal Society B, 279,* 1691–1699.

Romano, P. G., Horton, P., & Gray, J. E. (2004). The *Arabidopsis* cyclophilin gene family. *Plant Physiology, 134,* 1268–1282.

Rossi, L., Hohn, B., & Tinland, B. (1996). Integration of complete transferred DNA units is dependent on the activity of virulence E2 protein of *Agrobacterium tumefaciens. Proceedings of the National Academy of Sciences of the United States of America, 93,* 126–130.

Sahi, S. V., Chilton, M. D., & Chilton, W. S. (1990). Corn metabolites affect growth and virulence of *Agrobacterium tumefaciens. Proceedings of the National Academy of Sciences of the United States of America, 87,* 3879–3883.

Salomon, S., & Puchta, H. (1998). Capture of genomic and T-DNA sequences during double-strand break repair in somatic plant cells. *The EMBO Journal, 17,* 6086–6095.

Schrammeijer, B., Risseeuw, E., Pansegrau, W., Regensburg-Tuïnk, T. J. G., Crosby, W. L., & Hooykaas, P. J. J. (2001). Interaction of the virulence protein VirF of *Agrobacterium tumefaciens* with plant homologs of the yeast Skp1 protein. *Current Biology, 11,* 258–262.

Shi, Y., Lee, L. Y., & Gelvin, S. B. (2014). Is VIP1 important for *Agrobacterium*-mediated transformation? *The Plant Journal, 79,* 848–860.

Shilo, S., Tripathi, P., Melamed-Bessudo, C., Tzfadia, O., Muth, T. R., & Levy, A. A. (2017). T-DNA-genome junctions form early after infection and are influenced by the chromatin state of the host genome. *PLoS Genetics, 13,* e1006875.

Shimoda, N., Toyoda-Yamamoto, A., Aoki, S., & Machida, Y. (1993). Genetic evidence for an interaction between the VirA sensor protein and the ChvE sugar-binding protein of *Agrobacterium. The Journal of Biological Chemistry, 268,* 26552–26558.

Shimoda, N., Toyoda-Yamamoto, A., Nagamine, J., Usami, S., Katayama, M., Sakagami, Y., et al. (1990). Control of expression of *Agrobacterium vir* genes by synergistic actions of phenolic signal molecules and monosaccharides. *Proceedings of the National Academy of Sciences of the United States of America, 87,* 6684–6688.

Stachel, S. E., An, G., Flores, C., & Nester, E. W. (1985). A Tn3 *lacZ* transposon for the random generation of beta-galactosidase gene fusions: Application to the analysis of gene expression in *Agrobacterium. The EMBO Journal, 4,* 891–898.

Stachel, S. E., Messens, E., Van Montagu, M., & Zambryski, P. C. (1985). Identification of the signal molecules produced by wounded plant cell that activate T-DNA transfer in *Agrobacterium tumefaciens. Nature, 318,* 624–629.

Stachel, S. E., Timmerman, B., & Zambryski, P. C. (1986). Generation of single-stranded T-DNA molecules during the initial stages of T-DNA transfer for *Agrobacterium tumefaciens* to plant cells. *Nature, 322,* 706–712.

Stachel, S. E., & Zambryski, P. C. (1986). *virA* and *virG* control the plant-induced activation of the T-DNA transfer process of *A. tumefaciens. Cell, 46,* 325–333.

Steck, T. R., Morel, P., & Kado, C. I. (1988). Vir box sequences in *Agrobacterium tumefaciens* pTiC58 and A6. *Nucleic Acids Research, 16,* 8736.

Tao, Y., Rao, P. K., Bhattacharjee, S., & Gelvin, S. B. (2004). Expression of plant protein phosphatase 2C interferes with nuclear import of the *Agrobacterium* T-complex protein VirD2. *Proceedings of the National Academy of Sciences of the United States of America, 101,* 5164–5169.

Tinland, B., Schoumacher, F., Gloeckler, V., Bravo-Angel, A. M., & Hohn, B. (1995). The *Agrobacterium tumefaciens* virulence D2 protein is responsible for precise integration of T-DNA into the plant genome. *The EMBO Journal, 14,* 3585–3595.

Tiwari, M., Gautam, N., Indoliya, Y., Kidwai, M., Mishra, A. K., & Chakrabarty, D. (2022). A tau class GST, OsGSTU5, interacts with VirE2 and modulates the *Agrobacterium*-mediated transformation in rice. *Plant Cell Reports, 41*, 873–891.

Tzfira, T., Vaidya, M., & Citovsky, V. (2001). VIP1, an *Arabidopsis* protein that interacts with *Agrobacterium* VirE2, is involved in VirE2 nuclear import and *Agrobacterium* infectivity. *The EMBO Journal, 20*, 3596–3607.

Tzfira, T., Vaidya, M., & Citovsky, V. (2002). Increasing plant susceptibility to *Agrobacterium* infection by overexpression of the *Arabidopsis VIP1* gene. *Proceedings of the National Academy of Sciences of the United States of America, 99*, 10435–10440.

Tzfira, T., Vaidya, M., & Citovsky, V. (2004). Involvement of targeted proteolysis in plant genetic transformation by *Agrobacterium*. *Nature, 431*, 87–92.

Vaghchhipawala, Z. E., Vasudevan, B., Lee, S., Morsy, M. R., & Mysore, K. S. (2012). *Agrobacterium* may delay plant nonhomologous end-joining DNA repair via XRCC4 to favor T-DNA integration. *Plant Cell, 24*, 4110–4123.

van Attikum, H., Bundock, P., & Hooykaas, P. J. J. (2001). Non-homologous end-joining proteins are required for *Agrobacterium* T-DNA integration. *The EMBO Journal, 20*, 6550–6558.

van Attikum, H., Bundock, P., Overmeer, R. M., Lee, L. Y., Gelvin, S. B., & Hooykaas, P. J. J. (2003). The *Arabidopsis AtLIG4* gene is required for the repair of DNA damage, but not for the integration of *Agrobacterium* T-DNA. *Nucleic Acids Research, 31*, 4247–4255.

van Kregten, M., de Pater, S., Romeijn, R., van Schendel, R., Hooykaas, P. J. J., & Tijsterman, M. (2016). T-DNA integration in plants results from polymerase-theta-mediated DNA repair. *Nature Plants, 2*, 16164.

Vergunst, A. C., Schrammeijer, B., den Dulk-Ras, A., de Vlaam, C. M. T., Regensburg-Tuink, T. J., & Hooykaas, P. J. J. (2000). VirB/D4-dependent protein translocation from *Agrobacterium* into plant cells. *Science, 290*, 979–982.

Vergunst, A. C., van Lier, M. C. M., den Dulk-Ras, A., Stüve, T. A., Ouwehand, A., & Hooykaas, P. J. J. (2005). Positive charge is an important feature of the C-terminal transport signal of the VirB/D4-translocated proteins of *Agrobacterium*. *Proceedings of the National Academy of Sciences of the United States of America, 102*, 832–837.

Wang, Y., Haitjema, C. H., & Fuqua, C. (2014). The Ctp type IVb pilus locus of *Agrobacterium tumefaciens* directs formation of the common pili and contributes to reversible surface attachment. *Journal of Bacteriology, 196*, 2979–2988.

Wang, L., Lacroix, B., Guo, J., & Citovsky, V. (2017). Transcriptional activation of virulence genes of *Rhizobium etli*. *Journal of Bacteriology, 199*. e00841-16.

Wang, Y., Peng, W., Zhou, X., Huang, F., Shao, L., & Luo, M. (2014). The putative *Agrobacterium* transcriptional activator-like virulence protein VirD5 may target T-complex to prevent the degradation of coat proteins in the plant cell nucleus. *The New Phytologist, 203*, 1266–1281.

Wang, C., Ye, F., Chang, C., Liu, X., Wang, J., Wang, J., et al. (2019). Agrobacteria reprogram virulence gene expression by controlled release of host-conjugated signals. *Proçeedings of the National Academy of Sciences of the United States of America, 116*, 22331–22340.

Wang, L., Zhang, S., Huang, F., Zhou, X., Chen, Z., Peng, W., et al. (2018). VirD5 is required for efficient *Agrobacterium* infection and interacts with Arabidopsis VIP2. *The New Phytologist, 217*, 726–738.

Weisberg, A. J., Davis, E. W., 2nd, Tabima, J., Belcher, M. S., Miller, M., Kuo, C. H., et al. (2020). Unexpected conservation and global transmission of agrobacterial virulence plasmids. *Science, 368*. eaba5256.

Wendt, T., Doohan, F., & Mullins, E. (2012). Production of *Phytophthora infestans*-resistant potato (*Solanum tuberosum*) utilising *Ensifer adhaerens* OV14. *Transgenic Research, 21*, 567–578.

Willig, C. J., Duan, K., & Zhang, Z. J. (2018). Transcriptome profiling of plant genes in response to *Agrobacterium tumefaciens*-mediated transformation. *Current Topics in Microbiology and Immunology*, *418*, 319–348.

Wolterink-van Loo, S., Escamilla Ayala, A. A., Hooykaas, P. J. J., & van Heusden, G. P. (2015). Interaction of the *Agrobacterium tumefaciens* virulence protein VirD2 with histones. *Microbiology*, *161*, 401–410.

Wright, E. L., Deakin, W. J., & Shaw, C. H. (1998). A chemotaxis cluster from *Agrobacterium tumefaciens*. *Gene*, *220*, 83–89.

Xu, J., Kim, J., Danhorn, T., Merritt, P. M., & Fuqua, C. (2012). Phosphorus limitation increases attachment in *Agrobacterium tumefaciens* and reveals a conditional functional redundancy in adhesin biosynthesis. *Research in Microbiology*, *163*, 674–684.

Yadav, N. S., Vanderleyden, J., Bennett, D. R., Barnes, W. M., & Chilton, M. D. (1982). Short direct repeats flank the T-DNA on a nopaline Ti plasmid. *Proceedings of the National Academy of Sciences of the United States of America*, *79*, 6322–6326.

Yanofsky, M. F., Porter, S. G., Young, C., Albright, L. M., Gordon, M. P., & Nester, E. W. (1986). The *virD* operon of *Agrobacterium tumefaciens* encodes a site-specific endonuclease. *Cell*, *47*, 471–477.

Young, C., & Nester, E. W. (1988). Association of the VirD2 protein with the 5′ end of T-strands in *Agrobacterium tumefaciens*. *Journal of Bacteriology*, *170*, 3367–3374.

Yuan, Z. C., Edlind, M. P., Liu, P., Saenkham, P., Banta, L. M., Wise, A. A., et al. (2007). The plant signal salicylic acid shuts down expression of the *vir* regulon and activates quormone-quenching genes in *Agrobacterium*. *Proceedings of the National Academy of Sciences of the United States of America*, *104*, 11790–11795.

Zaltsman, A., Krichevsky, A., Loyter, A., & Citovsky, V. (2010). *Agrobacterium* induces expression of a plant host F-box protein required for tumorigenicity. *Cell Host & Microbe*, 7, 197–209.

Zhang, J., Boone, L., Kocz, R., Zhang, C., Binns, A. N., & Lynn, D. G. (2000). At the maize/*Agrobacterium* interface: Natural factors limiting host transformation. *Chemistry & Biology*, 7, 611–621.

Zhang, X., & Hooykaas, P. J. J. (2019). The *Agrobacterium* VirD5 protein hyper-activates the mitotic Aurora kinase in host cells. *The New Phytologist*, *222*, 1551–1560.

Zhang, X., van Heusden, G. P. H., & Hooykaas, P. J. J. (2017). Virulence protein VirD5 of *Agrobacterium tumefaciens* binds to kinetochores in host cells *via* an interaction with Spt4. *Proceedings of the National Academy of Sciences of the United States of America*, *114*, 10238–10243.

Zhu, Y., Nam, J., Carpita, N. C., Matthysse, A. G., & Gelvin, S. B. (2003). *Agrobacterium*-mediated root transformation is inhibited by mutation of an *Arabidopsis* cellulose synthase-like gene. *Plant Physiology*, *133*, 1000–1010.

Ziemienowicz, A., Görlich, D., Lanka, E., Hohn, B., & Rossi, L. (1999). Import of DNA into mammalian nuclei by proteins originating from a plant pathogenic bacterium. *Proceedings of the National Academy of Sciences of the United States of America*, *96*, 3729–3733.

Ziemienowicz, A., Tinland, B., Bryant, J., Gloeckler, V., & Hohn, B. (2000). Plant enzymes but not *Agrobacterium* VirD2 mediate T-DNA ligation *in vitro*. *Molecular and Cellular Biology*, *20*, 6317–6322.

Zipfel, C., Kunze, G., Chinchilla, D., Caniard, A., Jones, J. D., Boller, T., et al. (2006). Perception of the bacterial PAMP EF-Tu by the receptor EFR restricts *Agrobacterium*-mediated transformation. *Cell*, *125*, 749–760.

Zuniga-Soto, E., Mullins, E., & Dedicova, B. (2015). *Ensifer*-mediated transformation: An efficient non-*Agrobacterium* protocol for the genetic modification of rice. *Springerplus*, *4*, 600.

CHAPTER TWO

Epigenetic regulation of angiogenesis in tumor progression

Vanessa Desantis[a,†], Antonio G. Solimando[b,†], and Domenico Ribatti[c,*]

[a]Department of Biomedical Sciences and Human Oncology, Pharmacology Section, University of Bari Aldo Moro Medical School, Bari, Italy
[b]Department of Biomedical Sciences and Human Oncology (DIMO), Unit of Internal Medicine and Clinical Oncology, University of Bari Aldo Moro Medical School, Bari, Italy
[c]Department of Basic Medical Sciences, Neurosciences and Sensory Organs, University of Bari Medical School, Bari, Italy
*Corresponding author: e-mail address: domenico.ribatti@uniba.it

Contents

Abstract

Epigenetic is the study of those alterations regulating gene expression without altering DNA sequence and inherited by transmission through cell division. DNA hypomethylation, hypermethylation of tumor suppressor genes, aberrant histone modifications and/or specific microRNAs expression profiles contribute to tumor initiation and progression. In this review, we will discuss the role of epigenetic changes in the regulation of tumor angiogenesis.

† Equal contribution.

Advances in Genetics, Volume 110
ISSN 0065-2660
https://doi.org/10.1016/bs.adgen.2022.08.002

Abbreviations

3′-UTR	3′-untranslated region
CpG	cytosine-phosphate-guanine
EGFR	epidermal growth factor receptor
EGR1	early growth response 1
EHMT2	Euchromatic histone-lysine *N*-methyltransferase 2
EZH2	enhancer of zeste homolog 2
FGF-2	fibroblast growth factor-2
HATs	histone acetyltransferases
HDACs	histone deacetylases
HGF	hepatocyte growth factor
HIF	hypoxia inducible factor
IL-8	interleukin-8
MDB-2	Methyl-CpG-Binding Domain Protein-2
miRNAs	microRNAs
MM	multiple myeloma
MMP	matrix metalloproteinase
ROS	reactive oxygen specimens
RTK	receptor tyrosine kinase RTK
SAM	S-Adenosyl-methionine
SIRT-1	Sirtuin-1
TGFβ	transforming growth factor beta
TIMP	tissue inhibitor of matrix metalloproteinase
TSP-1	thrombospondin-1
VEGF	vascular endothelial growth factor
VEGFR	vascular endothelial growth factor receptor

1. Introduction

Conrad Waddington coined the term "epigenetics" to describe inherited changes in phenotype without changes in genotype; he said that "to understand how the genotypes of evolving organisms can respond to the environment in a more coordinated fashion" (Waddington, 1959). Epigenetic modifications play an integral role in the regulation of all DNA-based processes, including DNA replication, transcription, and repair. Cancer-related epigenome alterations are correlated with genetic mutations of chromatin (Shen & Laird, 2013), and non-genetic deregulations, including metabolic states, tumor microenvironment modifications, and aging.

Many factors influence epigenetic regulation, such as age, diet, lifestyle, environment, and disease, and accumulating evidence demonstrates that epigenetic modification changes during tumorigenesis. Epigenetic factors, including DNA methylation, histone modifications, chromatin remodeling and microRNAs (miRNAs or miRs) are involved in the regulation of tumor angiogenesis.

Angiogenesis, a multistage process of new capillaries formation from pre-existing vessels, is mediated by angiogenic stimulus released from different cytokines and growth factors. In physiological conditions, this process is essential for tissue maintenance and homeostasis during embryonic development and adult life, while in a pathological context angiogenesis is involved in tumor growth, progression and metastasis formation (Aspriţoiu, Stoica, Bleotu, & Diaconu, 2021; Ribatti, Vacca, Nico, Presta, & Roncali, 2003).

In this review, we summarized the role of epigenetic changes in the regulation of angiogenesis, inducing tumor progression.

2. DNA methylation

The genome of a tumor cell undergoes DNA hypermethylation or hypomethylation due to loss of DNA methylation in gene coding regions, introns, and repetitive DNA sequences (Feinberg & Vogelstein, 1983). A methyl group is added to the carbon-5-position of a cytosine in a cytosine-phosphate-guanine (CpG) dinucleotide, resulting in 5-methylcytosine (Smith & Meissner, 2013), Fig. 1.

Methylation abnormalities commonly occur in pathological disorders such as CpGs hypermethylation within the promoter regions of genes, inducing a silencing of these tumor suppressors (Aspriţoiu et al., 2021). The most common epigenetic changes in tumor cells include hypermethylation of anti-angiogenic factors, i.e., the silencing of the angiogenesis inhibitor thrombospondin-1 (TSP-1), that entail the secretion inhibition of transforming growth factor beta (TGFβ), supporting invasion and metastasis; the secretion of anti-angiogenic ADAM Metallopeptidase with Thrombospondin Type 1 Motif 8 (ADAMTS-8) and of tissue inhibitor of matrix metalloproteinases-2 and -9 (MMP-2 and MMP-3) (TIMP-2 and -3) (Buysschaert, Schmidt, Roncal, Carmeliet, & Lambrechts, 2008; Dunn et al., 2006; Hellström et al., 2007). Hypermethylation of the TSP-1 gene has been found in glioblastoma multiforme (Li, Ahuja, Burger, & Issa, 1999) and in neuroblastoma (Yang et al., 2003), it is also associated with poor prognosis, vascular invasion and metastasis in penile squamous carcinoma (Guerrero et al., 2008) and in gastric carcinoma (Miyamoto et al., 2007; Oue et al., 2003).

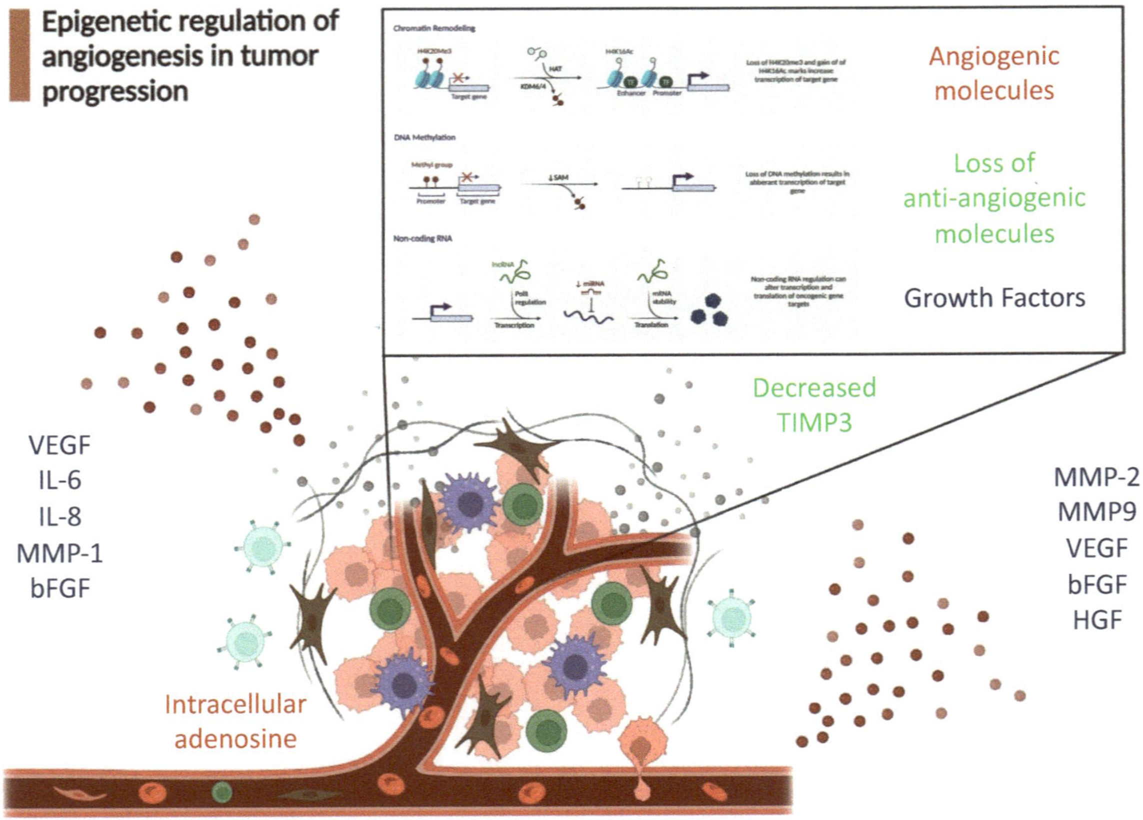

Fig. 1 Epigenetic regulation of angiogenesis in tumor progression.

Epigenetic reprogramming, in terms of DNA methylation, is also considered as a hallmark of senescence, impacting on DNA damage repair, telomere length, metabolic pathways and activation of senescence-related genes and miRNAs (Horvath, 2013). Specifically, Lyu et al. demonstrated that TGFβ/Smad signaling triggers miR-29-mediated reduction of H4K20me3 (the most common methylation state on histone H4) abundance, which promotes cellular senescence (Lyu et al., 2018).

Among positive factors that regulate angiogenesis, a great importance is associated to vascular endothelial growth factor (VEGF) and eNOS, down-regulated via Methyl-CpG-Binding Domain Protein-2 (MBD-2) (Rao et al., 2011). The loss of MBD-2 enhances angiogenesis with a higher ERK1/2 activity in endothelial cells, promoting apoptosis by increasing the expression of BCL-2. VEGF receptor-1 (VEGFR-1) and VEGFR-2 end their expression is controlled by promoter DNA methylation in cancer cells (Turunen & Ylä-Herttuala, 2011). DNA demethylation of VEGF-C is associated with gene expression and its possible involvement of lymphangiogenesis in gastric cancer (Matsumura, Oue, Mitani, Kitadai, & Yasui, 2007).

3. Histone modifications

Post-translational histone modifications include histone acetylation mediated by two different enzyme families: histone acetyltransferases (HATs) and histone deacetylases (HDACs). Eighteen HDACs enzymes have been identified in humans and separated into four classes. Overexpression of HDAC1 inhibits p63 and von Hippel-Lindau tumor suppressor gene expression and promotes angiogenesis under hypoxic conditions (Kim et al., 2001). A HDAC member expression, Sirtuin-1 (SIRT-1), increases deacetylation and is involved in emodin's anti-angiogenic activity in human osteosarcoma (Qu, Wang, Wu, Liu, & Hao, 2015, p. 1). Also, a higher expression of G9 histone methyltransferase in cervical cancer was associated with increased angiogenesis, while the inhibition of G9a (Euchromatic histone-lysine *N*-methyltransferase 2—EHMT2) is involved in cancer invasion and metastasis, decreasing angiogenic factor levels, including VEGF, interleukin-8 (IL-8), and angiogenin (Chen et al., 2018). SET-7, a histone methyltransferase, is up-regulated and associated with angiogenesis and poor prognosis in breast cancer (Zhang et al., 2016). High expression of the enhancer of zeste homolog 2 (EZH2) is associated with tumor angiogenesis, poor disease-free survival, and poor overall survival in intrahepatic cholangiocarcinoma (Nakagawa et al., 2018). HDAC inhibitors down-modulate angiogenic cytokine expression of tumor cells altering VEGF signaling and induce growth suppression and apoptosis in human mantle cell lymphoma (Heider et al., 2006).

4. The angiogenic role of miRNAs in tumor progression. Epigenetic regulation of angiogenesis in tumor progression

Epigenetic modification is crucial in a plethora of phenomena, comprising cancer vessel formation (Katoh, 2013; Subramaniam, Yehya, Cheng, Wang, & Oon, 2019). miRNAs are short 22-nucleotide noncoding RNA, which play a pivotal role in gene expression through sequence-specific base pairing in the 3′-untranslated region (3′-UTR) of the targeted RNA. It has been estimated that altered expression of miRNA has been reported in human cancers, regulating the expression of oncogenes and tumor suppressor genes, Fig. 1. Two types of miRNAs have been described in cancer, including tumor-suppressor miRNAs (anti-oncomiRs), and oncogenic miRNAs (oncomiRs), and their ectopic expression generate anti-tumor effects by inhibiting cell proliferation and increasing cell death (Desantis et al., 2020). It has been estimated that there are more than 1000 miRNA genes in the human genome, regulating more than one-third of the mRNA produced (Lewis, Burge, & Bartel, 2005).

miRNAs are important mediators of the interactions between tumor and endothelial cells. They modulate tumor angiogenesis through targeting pro- and anti-angiogenic factors, including receptor tyrosine kinase (RTK) signaling proteins, hypoxia inducible factor (HIF), VEGF, TSP-1, and reactive oxygen specimens (ROS) (Table 1).

The extracellular circulating miRNAs have been found to be encapsulated in lipid vesicles (30–150 nm diameter), including exosomes, which originate from early-endosome derived multivesicular bodies inside the cells, and are released through fusion with the plasma membrane (Denzer, Kleijmeer, Heijnen, Stoorvogel, & Geuze, 2000). Exosomal delivery of miRNAs to cancer cells may induce cancer progression and metastasis. miRNA-205 in exosomes secreted by ovarian cancer cells promote angiogenesis, tumor growth, and metastasis (He et al., 2019). miR-23a is upregulated in exosomes secreted by lung cancer under hypoxic conditions and increased angiogenesis and vascular permeability by targeting prolyl hydroxylase and tight junction protein ZO-1 (Hsu et al., 2017, p. 1). Glioma stem cells-derived exosome miR-26a promotes angiogenesis in glioma (Wang et al., 2019). miR-150-5p inhibits colorectal cancer tumor growth and angiogenesis by inactivating the VEGF-A/VEGFR-2 signaling pathway (Chen et al., 2018; Derakhshani et al., 2021) and miR-622-overexpression inhibited

Table 1 microRNAs acting as pro- and anti-angiogenic factors in tumors.

Role in angiogenesis	microRNAs	References
Pro-angiogenic RTKs signaling-related	miR-21; miR-126; miR-26a; miR-27-3p; miR-214-3p	Frassanito et al. (2019), Liu et al. (2011), Wang, Liao, Wu, and Dai (2019), and Zhang et al. (2013)
Anti-angiogenic RTKs signaling-related	miR-218; miR-18a	Saunders et al. (2010) and Zhang et al. (2017)
Anti-angiogenic HIF-targeted	miR-23a; miR-519c; miR-22; miR-107; miR-199a-5p	Hsu et al. (2017) and Raimondi et al. (2014)
Anti-angiogenic VEGF-targeted	miR-29b; miR-205; miR-15; miR-16; miR-7; miR-126; miR-150-5p;	Babae et al. (2014), Chen et al. (2018), Fang, Sun, Wang, and Wang (2019), Szczyrba et al. (2013), and Yue et al. (2012)
Pro-angiogenic TSP-1-targeted	miR-194; miR-17-92; miR-467	Gajeton et al. (2021), Sundaram et al. (2011), and Yang et al. (2016)
Pro-angiogenic ROS-signaling related	mi-R21	Yan et al. (2018), p. 21

colon cancer cell angiogenesis by suppressing the CXCR4-VEGF-A signaling pathway (Fang et al., 2019).

In prostate cancer cell lines, miR-29b was significantly reduced, leading to elevated VEGF-A levels (Szczyrba et al., 2013). miR-205 functions as a tumor suppressor in human glioblastoma cells by targeting VEGF-A (Yue et al., 2012). miR-15 and miR-16 regulate the expression of VEGF by inhibiting angiogenesis in different tumors (Lages et al., 2012). Systemic miR-7 delivery inhibits tumor angiogenesis and growth in murine xenograft glioblastoma (Babae et al., 2014). miR-192 is an anti-angiogenic miRNA acting through the inhibition of different pro-angiogenic pathways by regulating homeobox B9 and early growth response 1 (EGR1) (Wu et al., 2016). The epigenetic silencing of miR-126 led to upregulation of VEGF expression and contributed to tumor invasion and angiogenesis in colorectal cancer (Zhang et al., 2013).

In hematological malignancies, tumor-derived exosomes suppress antitumor immunity, promoting angiogenesis and drug resistance. Exosomes play a crucial role in remodeling the bone marrow microenvironment and multiple myeloma (MM) progression by transporting DNA, miRNA, and proteins. Roccaro et al. demonstrated that MM exosomes are angiogenic and modulate multiple signaling pathways in endothelial cells and bone marrow stromal cells to promote their growth (Roccaro et al., 2013). Moreover, in MM, cell-derived exosomes induce overexpression of miR-27-3p and miR-214-3p in fibroblasts, stimulating angiogenesis and tumor growth (Frassanito et al., 2019) while miR-199a-5p downregulates HIF-1α and the expression of angiogenic cytokines, including VEGF, IL-8, hepatocyte growth factor (HGF) and fibroblast growth factor-2 (FGF-2) (Raimondi et al., 2014), also shaping invasiveness and cell adhesion and mediating intra- and extramedullary dissemination (Da Vià et al., 2019; Solimando, Annese, et al., 2020; Solimando, Da Vià, et al., 2020).

5. Epigenetic inhibitors as anti-angiogenic agents in cancer treatment

DNA methyltransferase inhibitors are used to revert aberrant DNA methylation patterns in cancer cells. S-Adenosyl-methionine (SAM) modifies the gene expression and methylation providing methyl groups (Wang, Sun, & Szyf, 2017). Different studies have demonstrated a wide utility of treating cancer cells with histone- and DNA-methylation-modifying agents, so-called epigenetic therapy. One of the earliest mechanisms of action for epigenetic therapy was believed to be on changing the transcription profiles, although, at high doses, demethylating agents have been shown to exhibit cytotoxic effects, which result in generalized cellular death. However, under conditions where prolonged exposure to lower doses of epigenetic therapies has been conducted, cellular reprogramming through changes in transcription has been observed. A tumor, which might initially respond to conventional chemotherapy, but no longer responds, becomes sensitive to the chemotherapeutic agents once again in the presence of these epigenetic modifiers. Solid tumors are targets of DNA demethylating therapy and different clinical trials are being conducted, focused on a combination of DNA demethylating agents and conventional cytotoxic drugs. A combination of decitabine, a hypomethylating agent, and cisplatin was effective for advanced cervical cancers (Pohlmann et al., 2002) and decitabine and an anti-epidermal growth factor receptor (EGFR) antibody was effective for metastatic colorectal cancer with wild type KRAS (Garrido-Laguna et al., 2013). Studies using decitabine have demonstrated the upregulation of *p16*, which in turn downregulates VEGF, as pro-angiogenic factor (Miki, Shimizu, Yano, Tani, & Sone, 2001).

The reversibility of epigenetic events makes them suitable for therapeutic intervention. The anti-angiogenic effects of HDAC inhibitors are associated with downregulation of pro-angiogenic genes, including VEGF and/or eNOS. Trichostatin A (TSA), a selective inhibitor of HDACs inhibits HIF-1α and VEGF expression in human tongue squamous cell carcinoma cells in vitro, increasing the expression of p53 in Lewis lung carcinoma in vivo models (Kang, Que, Wu, Wang, & Sun, 2012; Kim et al., 2001), suggesting that HDAC inhibitor have a potential therapeutic value for cancer therapy.

Inhibition of HIF-1α by HDACi provides a rationale for combining HDACi with other anti-angiogenic agents such as VEGFRs tyrosine kinase inhibitors by targeting multiple pathways in tumor progression and angiogenesis (Qian et al., 2004, p. 222584).

Another inhibitor of HDACs is superpolyamide hydroxamic acid (SAHA) that in combination with TSA, inhibits VEGFR-1 and VEGFR-2 and their co-receptors neuropilin-1 and -2 (NRP-1 and NRP-2) expression in lung cancer cell lines (Marks, 2007). In breast cancer, SAHA reduces VEGF-C expression in a dose-dependent manner (Cheng & Hung, 2013) and in human mantle cell lymphoma downregulates VEGF by an unknown mechanism (Heider et al., 2006).

5-Azacytidine is effective in 50–60% of patients with myelodysplastic syndrome, inhibiting antiangiogenic agents by direct methylation to the promoter of TSP-1, JUNB, and IGFBP3 genes, which are suppressed in tumor-conditioned endothelial cells (Hellebrekers et al., 2006; Raj & Mufti, 2006). RRx-001, an epigenetic-based radio and chemosensitizer can induce a transient normalization of tumor vasculature to make it more efficient for oxygen and drug delivery (Oronsky, Scicinski, Cabrales, & Minchinton, 2016).

Clinical trials with demethylating agents and HDAC inhibitors have been employed against hematological tumors. Two pan-HDAC inhibitors, Vorinostat and Romidepsin, have been used for the treatment of cutaneous T cell lymphoma (Olsen et al., 2007; Piekarz et al., 2009).

The use of miRNAs in the therapeutic targeting of tumor angiogenesis has been obtained through the delivery of antago-miRNA or miRNA mimics both in tumor cells and in tumor stromal cells (Frassanito et al., 2019; Garzon, Marcucci, & Croce, 2010; Saltarella et al., 2022). The intracellular delivery of anti-angiogenic miRNAs with low toxicity has been obtained using liposomes with acetylated polyethylenimine (PEI, polycation liposomes) (Muramatsu, Kidoya, Naito, Sakimoto, & Takakura, 2013).

6. The hematological malignancies roadmap

6.1 Gene expression profiling (GEP) and epigenetic regulation

Neoplasms of hematopoietic tissue originate from the accumulation of multiple genetic and epigenetic aberrations acquired at the somatic level in hematopoietic stem/progenitor cells, resulting in an altered gene expression. Epigenetic modifications can influence gene expression, inducing deleterious effects that can result in cell malignant transformations (Heuck et al., 2013). The identification of specific cytogenetic and molecular alterations is the basis of the current classification of hematopoietic tissue neoplasms and is of considerable importance in the treatment planning and prognostic stratification of patients.

At the beginning of the 2000s, the advent of new technologies in the study of the genome allowed a better characterization of the molecular alterations underlying the clinical and biological heterogeneity of myeloid and lymphoid neoplasms. GEP based on microarray technology, evaluating the simultaneous expression of thousands of genes, has identified specific gene expression profiles associated with distinct diseases. In particular, the GEP made it possible to identify characteristic patterns of gene expression in subclasses of myeloid and lymphoid neoplasms with specific genetic and molecular aberrations, to identify new subclasses of leukemia and predict the response to standard therapy, simultaneously offering new insights into the pathogenesis of neoplasms and identifying new molecular targets for the discovery of new drugs (Theilgaard-Mönch et al., 2011).

Although different technologies and microarray platforms have been used, there is strong evidence that GEP technology can be standardized for diagnostics as distinct gene expression patterns are very robust with high reproducibility both within the same platform and between different platforms.

In MM GEP analyses have been used to define the risk, to classify the disease stages, to identify drug resistance and therapeutic targets and consequently to predict the treatment response (Chng et al., 2016). Since in MM great importance is associated with the interaction between MM plasma cells and the bone marrow microenvironment (Gnoni et al., 2020; Leone et al., 2020; Solimando et al., 2021), which in turn controls MM plasma cells growth, survival and drug resistance (Saltarella et al., 2020), GEP approaches can also be used to study this complex interaction (Mahtouk et al., 2010).

The Microarray Innovations in Leukemia (MILE) collaborative study program born around the European LeukemiaNet international study group comprising 11 laboratories from three different continents, evaluated the clinical accuracy of gene expression profiles, compared with the current routine diagnostic algorithm, for 16 subclasses of acute and chronic leukemias, myelodysplastic syndromes and a control group including normal and non-malignant spinal cord disorders. In a first cohort of 2096 patients, the GEP achieved a diagnostic accuracy of 92.2% for 18 distinct disease subtypes with a median specificity of 99.7%. In the second cohort of 1152 prospectively collected patients, the GEP classification achieved a median sensitivity of 99.7% and a median specificity of 99.8% for 14 standard acute leukemia subtypes (Haferlach et al., 2010).

The GEP can predict the presence of recurrent genetic alterations in patients with Acute Myeloid Leukemia, such as t (8; 21), inv16, t (15; 17) and the alterations of the mixed lineage leukemia (MLL)/t gene (11q23), and of molecular alterations such as mutations in the transcription factor CCAAT/enhancer binding protein α (CEBPA) and in NPM1. In addition, through GEP studies, a panel of 133 genes capable of predicting patient prognosis was identified (Mills et al., 2009).

GEP studies in patients with myelodysplastic syndromes were performed both on purified $CD34^+$ or $CD133^+$ cells, and on unfractionated bone marrow mononuclear cells, with more heterogeneous results than in acute leukemia studies. In general, GEP studies have commonly shown deregulation of genetic pathways associated with the signaling of Interferon and thrombopoietin and the Wnt pathway. Low-risk forms of myelodysplastic syndromes show deregulation of immune surveillance genes, apoptosis and chemokine signaling pathways, while the high-risk forms are characterized by impaired response to DNA damage and cell cycle checkpoints. Distinct genetic expression profiles have been described in the forms with recurrent genetic alterations such as 5q-, trisomy 8 and 7- / del (7q). In the MILE study, the diagnostic accuracy of GEP was however only 50% on unfractionated bone marrow mononuclear cell samples from patients with myelodysplastic syndrome but allowed the identification of a gene profile with prognostic value associated with an increased risk of transformation into Acute Leukemia (Theilgaard-Mönch et al., 2011).

The GEP studies have allowed the identification of two main molecular subtypes in the context of Diffuse Large B Cell Lymphomas (DLBCL): the first characterized by the expression of a set of genes typical of the B lymphocyte of the germinal center is called germinal center B-cell like (GCB) DLBCL; the other subtype, called activated B-cell like (ABC) DLBCL, expresses a set of up-regulated genes in activated peripheral blood B lymphocytes and has a worse prognosis than GCB-DLBCL. A small subset of patients exhibit a different gene expression profile from both GCB- and ABC-DLBCL and are referred to as unclassifiable (Alizadeh et al., 2000; Solimando, Annese, et al., 2020; Solimando, Da Vià, et al., 2020). Primary mediastinal cell lymphoma has different molecular characteristics from GCB- and ABC-DLBCL, with enrichment in genes involved in the IL13 signaling pathway, molecularly resembling Hodgkin's lymphoma (Savage et al., 2003).

The use of the GEP is currently reserved only for clinical studies and has not yet entered the routine diagnostic algorithm of hematopoietic tissue neoplasms. Although it has enormous potential in identifying multiple genetic and molecular alterations imported for the classification and prognostic stratification of leukemia, it is a constantly evolving technology that requires significant technical and analytical support. Furthermore, further studies are needed to give the right meaning to the huge amount of data obtained with the analysis of the GEP.

6.2 Next generation sequencing (NGS) and epigenetic regulation

DNA sequencing by NGS or deep sequencing represents the main innovation in molecular diagnostics in onco-hematology in recent years. Compared to traditional sequencing, NGS allows the massive and parallel sequencing of DNA fragments that cover large portions of the genome if not the entire genome, with greater accuracy, fewer sequence errors, reduced times, and lower costs than traditional sequencing. There has been a rapid proliferation in the number of NGS platforms available on the market, including Illumina, Applied Biosystems SOLiD System, 454 Life Sciences (Roche), Helicos HeliScope, Complete Genomics, Pacific Biosciences PacBio, Life Technologies Ion Torrent and Oxford Nanopore.

The applications of NGS to the study of neoplasms of hematopoietic and lymphoid tissue include the sequencing of the whole genome (Whole-Genome Sequencing, WGS), of the coding regions only (Whole-Exome Sequencing, WES) of a small number of genes of importance for the disease under study (targeted-NSG), or of the transcriptome (RNA-seq). NSG can also be applied to the study of epigenetic markers and chromatin structure of the entire genome using alternative sequence methods such as ChIP-seq, methyl-seq and DNase-seq.

The WGS offers the possibility of identifying a complete range of genomic alterations including point mutations, insertions/deletions (indels), changes in copy number, structural rearrangements such as translocations, cryptic rearrangements, inversions and complex rearrangements of the entire genome. The sequencing depth coverage is 20–40 ×. The pathological specimen and the patient's normal specimen are generally analyzed in combination to distinguish between polymorphism and somatic mutation.

WES is a cheaper method that aims to identify mutations only in the coding sequences for proteins, equal to approximately 1% of the genome, with a consequent reduction in the enormous amount of data to be analyzed and time. WES includes an enrichment step of genomic DNA for exons (exome-capture) through antisense oliconucleotides and separation with magnetic beads. Given that 85% of mutations fall into the coding regions, WES represents a valid alternative to WGS. The disadvantages are represented by the inability to identify chromosomal translocations, which often involve intronic regions, and structural alterations such as large deletions or insertions. The depth of sequencing yield is 100 ×, ensuring greater sensitivity than WGS for the search for low-frequency somatic mutations.

RNA-seq or transcriptome sequence is a technique that allows you to sequence transcribed, coding and non-coding RNA (ncRNA) genes. Compared to WGS, RNA-seq gives information on the level of gene expression, identifies expressed fusion transcripts and post-transcriptional modifications such as alternative splicing, single nucleotide variants (SNV) and RNA editing events.

Targeted-NGS consists in the sequencing of a small panel of genes significant for the pathology under study, after enriching the genomic DNA for target sequences, with a significant reduction in costs, times and quantity of data produced at the end of the process, with greater utility in the clinical context. This approach further improves the depth of sequencing yield to 500–1000 ×, allowing detection of allele variants below 2%.

The steps of NGS are similar between the various methods with some variations depending on the platform used and include template preparation, nucleic acid sequencing, imaging and data analysis. Template preparation consists of building and amplifying a nucleic acid library (genomic DNA or cDNA), cutting the DNA sample into fragments of approximately

500 bp or less and binding the adapter sequences to the ends of the fragments. The libraries are then amplified before sequencing by emulsion PCR or solid phase amplification depending on the method used and subjected to sequencing. In some technologies, sequencing takes place by synthesis using the fragments of the library as a template for the synthesis of cDNA fragments and digital recording of the nucleotides inserted by means of fluorometric or ion-sensitive methods (semiconductor-sequencing). The data obtained is processed to remove adapters, eliminate low-quality reads, align to the reference sequence or reconstruct the sequence, and analyze the complete sequence.

The advent of NSG has allowed the identification of numerous new genetic lesions and potential new therapeutic targets, which have helped to define and better the pathogenesis, classification and prognosis of various myeloid and lymphoid neoplasms, demonstrating how a molecular approach based on the study of a single gene does not reflect the complexity of the pathology and is unable to provide the clinician with all prognostic information deriving from the biology of the disease.

In acute myeloid leukemia (AML) NGS has led to the discovery of new genetic lesions, especially in patients with normal karyotype, including mutations in the genes DNMT3A, IDH1, IDH2, TET2, ASXL1, which have shown prognostic significance in several clinical studies. Mutations in the DNMT3A gene are present in 29–36% of patients with normal karyotype acute myeloid leukemia (CN-AML), mainly represented by missense mutations in the amino acid R882, inactivate the DNMT3A enzyme and have been associated with lower OS and RFS. Recurrent mutations of the genes for the isocitrate dehydrogenase isoforms IDH1 and IDH2 are found in 16% of the CN-AMLs, they concern the codons R132 of the IDH1 and R140 or R172 of the IDH2, resulting in an increase in the activity of these enzymes with increased transformation of alpha-ketoglutarate to 2-hydroxyglutarate, and are associated with a poor prognosis in terms of OS and DFS in patients with normal karyotype. Mutations of the TET2 gene, involved in DNA demethylation, are found in 24% of AML, but also in other myeloid neoplasms, and are associated with a poor prognosis in patients with intermediate karyotype. Mutations in the ASXL1 gene, which codes for a protein of the Polycomb group, are found in 5–10% of AMLs, especially in patients with intermediate karyotype, and are associated with lower OS (Ilyas et al., 2015; White & DiPersio, 2014).

The advent of new high-throughput sequencing methods has led to the identification of genetic lesions in 80–90% of myelodysplastic syndromes

(MDS), compared to 20–30% of conventional cytogenetics, significantly increasing the knowledge on the molecular pathogenesis of these diseases and prognostic stratification of patients. Genes commonly mutated in MDS show significant overlap with those mutated in AML and can be divided into five functional categories, including genes involved in the spliceosome (U2AF1, SF3B1, SRSF2, and ZRSR2), epigenetic modifiers (TET2, DNMT3A, BCOR, ASXL1, IDH1, and IDH2), cohins (STAG2, RAD21, and SMC3), transcription factors (RUNX1, WT1, and ETV6), and signaling molecules (NF1, NRAS, CBL, PTPN11, JAK2, and FLT3). Mutations of SF3B1 are more frequent in patients with refractory anemia with ring sideroblasts (RARS) and refractory cytopenia with multilinear dysplasia and ring sideroblasts (RCMD-RS), and confer a favorable prognostic significance (Malcovati et al., 2013; Solimando, Melaccio, Vacca, & Ria, 2022). Mutations in the TP53, EZH2, ETV6, RUNX1, and ASXL1 genes have been associated with reduced survival in patients with MDS, independently of IPSS. Several targeted-NGS panels have been developed and validated on large patient cohorts. From mutational analysis of 104 by deep-sequencing on 944 patients with various MDS subtypes, Haferlach et al. found mutations in 90% of patients, with a median frequency of three mutations per patient. Forty-six genes were significantly mutated with 6 genes mutated in more than 10% of patients (TET2, SF3B1, ASXL1, SRSF2, DNMT3A, and RUNX1). The mutational status of 14 genes in association with conventional IPSS-R prognostic factors was used to define a new prognostic model that discriminates 4 risk groups (low, intermediate, high and very high) with 3-year survival of 95.2, 69.3, 32.8 and 5.3% ($P < 0.001$) (Haferlach et al., 2014).

The application of NGS techniques has documented pathogenetically relevant recurrent genetic lesions in multiple lymphoproliferative diseases, including chronic lymphocytic leukemia, diffuse large B cell lymphoma, Burkitt's lymphoma, marginal splenic lymphoma, mantle cell lymphoma, leukemia with hairy cells and Waldenström's Macroglobulinemia (Shi et al., 2007; Solimando, Ribatti, Vacca, & Einsele, 2016).

WES applied to Chronic Lymphocytic Leukemia has led to the identification of mutations in the NOTCH1, BIRC3 and SF3B1 genes which, although uncommon in newly diagnosed patients, are significantly more represented in patients with chemo-refractory disease and in patients with transformations into Richter. NGS is also able to identify mutations and deletions of the TP53 and ATM genes, with higher sensitivity than Sanger sequencing and FISH, and allows the identification of subclones with a significant impact on prognosis. In consideration of the relatively low genetic complexity of Chronic Lymphocytic Leukemia, a targeted NGS approach is desirable for a correct stratification of patient risk and consequent better therapeutic choice, average by identifying mutations and deletions of the TP53 and ATM genes, mutational status and stereotypy of the IGVH genes, and mutations of the NOTCH1m BIRC3 and SF3B1 genes (Zent & Burack, 2014, p. 53).

In Waldenstrom Macroglobulinemia (WM), WGS identified recurrent activating mutations of the MYD88 L265P and CXCR4 genes in 90% and 27% of affected patients, respectively. These mutations are associated with the clinical characteristics of patients and have an impact on prognosis. MYD88 L265P mutations have already been described above. Mutations of CXCR4 are nonsense or frameshift mutations similar to those observed at the germline level in WHIM syndrome (warts, hypogammaglobulinemia, infections and myelokathexis), and lead to a truncated and constitutively active form of the receptor of the intracellular portion. The mutational status of the MYD88 and CXCR4 genes influences the response of WM patients to Ibrutinib, with ORR of 100% in MYD88 (L265P) patients CXCR4 (WT), 85% in MYD88 (L265P) CXCR4 (WHIM) patients, and 71.4% in MYD88 (WT) CXCR4 (WT) patients (Treon, Xu, & Hunter, 2015).

NGS methods applied to diffuse large B-cell lymphomas (DLBCL) revealed a complex genetic heterogeneity characterized mainly by mutations of genes that play crucial roles in chromatin modification, in signaling pathways (BCR signaling, Toll-like receptor signaling, NOTCH1 signaling, NF-κB pathway and PI3K pathway), in the control of the cell cycle and apoptosis. DLBCL with activated B cell-type (ABC) phenotype are characterized by MYD88 mutations, activator of the NF-κB and JAK-STAT3 pathways following stimulation of the Toll-like receptor and interleukin receptors, in about 30% of cases (L265P substitution more common), and of genes for CD79A and CD79B, involved in BCR signaling, in 18% of cases. DLBCL with germinal-center B cell-like phenotype, as well as follicular lymphomas, are characterized by mutations of EZH2 (in 23% of DLBCL GCB and 11% of FL), MEP2B (in 12% of DLBCL GCB and 11% of FL), BCL2, GNA13 and SGK1. Common to both DLBCL subtypes are mutations of CARD11, activator of NF-κB (in 10% of DLBCL), of CREBBP and EP300, encoding for histone-acetyltransferase (in 32% of GCB and 13% of non-GBC), of MLL2, encoding histone methyltransferase (in 33% of DLBCL, but also present in 90% of follicular lymphomas) and FOXO1 (in 8.6% of DLBCL). Despite the significant molecular heterogeneity, these mutated genes have been used as potential targets of target therapies and their prognostic significance has been validated in clinical trials with the use of targeted-NSG (Jardin, 2014).

The use of deep sequencing methods applied to the genes of the variable regions of Immunoglobulins (IgVH) and of the T-cell receptor (TCR) has also been applied to the monitoring of minimal residual disease after therapy in acute lymphoblastic leukemias, lymphomas and MM, revealing a higher sensitivity than conventional flow cytometry and ASO-PCR methods (Rigolin et al., 2016; Solimando et al., 2019).

NGS has also been used on circulating-tumor DNA (ctDNA) to search for tumor-specific mutations and rearrangements of genes for IgVH and TCR. The ctDNA has a short half-life (about 2 h) which allows, compared to other biomarkers, a more current evaluation of the state of the tumor and appears in the blood of patient's weeks if not months before the visualization of the tumor with the diagnostic and imaging methods, representing an ideal sample for molecular monitoring of patients with higher sensitivity than whole blood.

7. Conclusions

In conclusion, a molecular-genetic profile using deep-sequencing represents a feasible and highly promising approach for a better understanding of the pathogenesis and greater diagnostic accuracy, biological subclassification and prognostic stratification of patients with hematopoietic and lymphoid tissue neoplasms. Considering the progressive reduction of costs and therefore the availability of this technology for the study of large numbers of patients, in the very near future the molecular profile using target-NGS will have to be integrated with the current morphological, flow cytometric and genetic-molecular methods in the diagnostic process of hematological neoplasms with a significant impact on therapeutic decisions.

Finally, NGS allows the detection of somatic mutations in peripheral blood or bone marrow even in normal individuals, defining the presence of clonal hematopoiesis.

References

Alizadeh, A. A., Eisen, M. B., Davis, R. E., Ma, C., Lossos, I. S., Rosenwald, A., et al. (2000). Distinct types of diffuse large B-cell lymphoma identified by gene expression profiling. *Nature*, *403*, 503–511. https://doi.org/10.1038/35000501.

Aspriţoiu, V. M., Stoica, I., Bleotu, C., & Diaconu, C. C. (2021). Epigenetic regulation of angiogenesis in development and tumors progression: Potential implications for cancer treatment. *Frontiers in Cell and Development Biology*, *9*, 689962. https://doi.org/10.3389/fcell.2021.689962.

Babae, N., Bourajjaj, M., Liu, Y., Van Beijnum, J. R., Cerisoli, F., Scaria, P. V., et al. (2014). Systemic miRNA-7 delivery inhibits tumor angiogenesis and growth in murine xenograft glioblastoma. *Oncotarget*, *5*, 6687–6700. https://doi.org/10.18632/oncotarget.2235.

Buysschaert, I., Schmidt, T., Roncal, C., Carmeliet, P., & Lambrechts, D. (2008). Genetics, epigenetics and pharmaco-(epi)genomics in angiogenesis. *Journal of Cellular and Molecular Medicine*, *12*, 2533–2551. https://doi.org/10.1111/j.1582-4934.2008.00515.x.

Chen, X., Xu, X., Pan, B., Zeng, K., Xu, M., Liu, X., et al. (2018). miR-150-5p suppresses tumor progression by targeting VEGFA in colorectal cancer. *Aging (Albany NY)*, *10*, 3421–3437. https://doi.org/10.18632/aging.101656.

Cheng, H.-T., & Hung, W.-C. (2013). Inhibition of lymphangiogenic factor VEGF-C expression and production by the histone deacetylase inhibitor suberoylanilide hydroxamic acid in breast cancer cells. *Oncology Reports*, *29*, 1238–1244. https://doi.org/10.3892/or.2012.2188.

Chng, W. J., Chung, T.-H., Kumar, S., Usmani, S., Munshi, N., Avet-Loiseau, H., et al. (2016). Gene signature combinations improve prognostic stratification of multiple myeloma patients. *Leukemia*, *30*, 1071–1078. https://doi.org/10.1038/leu.2015.341.

Da Vià, M. C., Solimando, A. G., Garitano-Trojaola, A., Barrio, S., Munawar, U., Strifler, S., et al. (2019). CIC mutation as a molecular mechanism of acquired resistance to combined BRAF-MEK inhibition in extramedullary multiple myeloma with central nervous system involvement. *The Oncologist*. https://doi.org/10.1634/theoncologist.2019-0356.

Denzer, K., Kleijmeer, M. J., Heijnen, H. F., Stoorvogel, W., & Geuze, H. J. (2000). Exosome: From internal vesicle of the multivesicular body to intercellular signaling device. *Journal of Cell Science*, *113*(Pt 19), 3365–3374. https://doi.org/10.1242/jcs.113.19.3365.

Derakhshani, A., Hashemzadeh, S., Asadzadeh, Z., Shadbad, M. A., Rasibonab, F., Safarpour, H., et al. (2021). Cytotoxic T-lymphocyte antigen-4 in colorectal cancer: Another therapeutic side of capecitabine. *Cancers (Basel)*, *13*, 2414. https://doi.org/10.3390/cancers13102414.

Desantis, V., Saltarella, I., Lamanuzzi, A., Melaccio, A., Solimando, A. G., Mariggiò, M. A., et al. (2020). MicroRNAs-based nano-strategies as new therapeutic approach in multiple myeloma to overcome disease progression and drug resistance. *International Journal of Molecular Sciences*, *21*, E3084. https://doi.org/10.3390/ijms21093084.

Dunn, J. R., Reed, J. E., du Plessis, D. G., Shaw, E. J., Reeves, P., Gee, A. L., et al. (2006). Expression of ADAMTS-8, a secreted protease with antiangiogenic properties, is downregulated in brain tumours. *British Journal of Cancer*, *94*, 1186–1193. https://doi.org/10.1038/sj.bjc.6603006.

Fang, Y., Sun, B., Wang, J., & Wang, Y. (2019). miR-622 inhibits angiogenesis by suppressing the CXCR4-VEGFA axis in colorectal cancer. *Gene*, *699*, 37–42. https://doi.org/10.1016/j.gene.2019.03.004.

Feinberg, A. P., & Vogelstein, B. (1983). Hypomethylation distinguishes genes of some human cancers from their normal counterparts. *Nature*, *301*, 89–92. https://doi.org/10.1038/301089a0.

Frassanito, M. A., Desantis, V., Di Marzo, L., Craparotta, I., Beltrame, L., Marchini, S., et al. (2019). Bone marrow fibroblasts overexpress miR-27b and miR-214 in step with multiple myeloma progression, dependent on tumour cell-derived exosomes. *The Journal of Pathology*, *247*, 241–253. https://doi.org/10.1002/path.5187.

Gajeton, J., Krukovets, I., Muppala, S., Verbovetskiy, D., Zhang, J., & Stenina-Adognravi, O. (2021). Hyperglycemia-induced miR-467 drives tumor inflammation and growth in breast Cancer. *Cancers (Basel)*, *13*, 1346. https://doi.org/10.3390/cancers13061346.

Garrido-Laguna, I., McGregor, K. A., Wade, M., Weis, J., Gilcrease, W., Burr, L., et al. (2013). A phase I/II study of decitabine in combination with panitumumab in patients with wild-type (wt) KRAS metastatic colorectal cancer. *Investigational New Drugs, 31*, 1257–1264. https://doi.org/10.1007/s10637-013-9947-6.

Garzon, R., Marcucci, G., & Croce, C. M. (2010). Targeting microRNAs in cancer: Rationale, strategies and challenges. *Nature Reviews. Drug Discovery, 9*, 775–789. https://doi.org/10.1038/nrd3179.

Gnoni, A., Brunetti, O., Longo, V., Calabrese, A., Argentiero, A., Calbi, R., et al. (2020). Immune system and bone microenvironment: Rationale for targeted cancer therapies. *Oncotarget, 11*, 480–487. https://doi.org/10.18632/oncotarget.27439.

Guerrero, D., Guarch, R., Ojer, A., Casas, J. M., Ropero, S., Mancha, A., et al. (2008). Hypermethylation of the thrombospondin-1 gene is associated with poor prognosis in penile squamous cell carcinoma. *BJU International, 102*, 747–755. https://doi.org/10.1111/j.1464-410X.2008.07603.x.

Haferlach, T., Kohlmann, A., Wieczorek, L., Basso, G., Kronnie, G. T., Béné, M.-C., et al. (2010). Clinical utility of microarray-based gene expression profiling in the diagnosis and subclassification of leukemia: Report from the international microarray innovations in leukemia study group. *Journal of Clinical Oncology, 28*, 2529–2537. https://doi.org/10.1200/JCO.2009.23.4732.

Haferlach, T., Nagata, Y., Grossmann, V., Okuno, Y., Bacher, U., Nagae, G., et al. (2014). Landscape of genetic lesions in 944 patients with myelodysplastic syndromes. *Leukemia, 28*, 241–247. https://doi.org/10.1038/leu.2013.336.

He, L., Zhu, W., Chen, Q., Yuan, Y., Wang, Y., Wang, J., et al. (2019). Ovarian cancer cell-secreted exosomal miR-205 promotes metastasis by inducing angiogenesis. *Theranostics, 9*, 8206–8220. https://doi.org/10.7150/thno.37455.

Heider, U., Kaiser, M., Sterz, J., Zavrski, I., Jakob, C., Fleissner, C., et al. (2006). Histone deacetylase inhibitors reduce VEGF production and induce growth suppression and apoptosis in human mantle cell lymphoma. *European Journal of Haematology, 76*, 42–50. https://doi.org/10.1111/j.1600-0609.2005.00546.x.

Hellebrekers, D. M. E. I., Jair, K.-W., Viré, E., Eguchi, S., Hoebers, N. T. H., Fraga, M. F., et al. (2006). Angiostatic activity of DNA methyltransferase inhibitors. *Molecular Cancer Therapeutics, 5*, 467–475. https://doi.org/10.1158/1535-7163.MCT-05-0417.

Hellström, M., Phng, L.-K., Hofmann, J. J., Wallgard, E., Coultas, L., Lindblom, P., et al. (2007). Dll4 signalling through Notch1 regulates formation of tip cells during angiogenesis. *Nature, 445*, 776–780. https://doi.org/10.1038/nature05571.

Heuck, C. J., Mehta, J., Bhagat, T., Gundabolu, K., Yu, Y., Khan, S., et al. (2013). Myeloma is characterized by stage-specific alterations in DNA methylation that occur early during myelomagenesis. *Journal of Immunology, 190*, 2966–2975. https://doi.org/10.4049/jimmunol.1202493.

Horvath, S. (2013). DNA methylation age of human tissues and cell types. *Genome Biology, 14*, R115. https://doi.org/10.1186/gb-2013-14-10-r115.

Hsu, Y.-L., Hung, J.-Y., Chang, W.-A., Lin, Y.-S., Pan, Y.-C., Tsai, P.-H., et al. (2017). Hypoxic lung cancer-secreted exosomal miR-23a increased angiogenesis and vascular permeability by targeting prolyl hydroxylase and tight junction protein ZO-1. *Oncogene, 36*, 4929–4942. https://doi.org/10.1038/onc.2017.105.

Ilyas, A. M., Ahmad, S., Faheem, M., Naseer, M. I., Kumosani, T. A., Al-Qahtani, M. H., et al. (2015). Next generation sequencing of acute myeloid leukemia: Influencing prognosis. *BMC Genomics, 16*(Suppl 1), S5. https://doi.org/10.1186/1471-2164-16-S1-S5.

Jardin, F. (2014). Next generation sequencing and the management of diffuse large B-cell lymphoma: From whole exome analysis to targeted therapy. *Discovery Medicine, 18*, 51–65.

Kang, F.-W., Que, L., Wu, M., Wang, Z.-L., & Sun, J. (2012). Effects of trichostatin A on HIF-1α and VEGF expression in human tongue squamous cell carcinoma cells in vitro. *Oncology Reports, 28*, 193–199. https://doi.org/10.3892/or.2012.1784.

Katoh, M. (2013). Therapeutics targeting angiogenesis: Genetics and epigenetics, extracellular miRNAs and signaling networks (review). *International Journal of Molecular Medicine, 32*, 763–767. https://doi.org/10.3892/ijmm.2013.1444.

Kim, M. S., Kwon, H. J., Lee, Y. M., Baek, J. H., Jang, J. E., Lee, S. W., et al. (2001). Histone deacetylases induce angiogenesis by negative regulation of tumor suppressor genes. *Nature Medicine, 7*, 437–443. https://doi.org/10.1038/86507.

Lages, E., Ipas, H., Guttin, A., Nesr, H., Berger, F., & Issartel, J.-P. (2012). MicroRNAs: Molecular features and role in cancer. *Frontiers in Bioscience (Landmark Edition), 17*, 2508–2540. https://doi.org/10.2741/4068.

Leone, P., Solimando, A. G., Malerba, E., Fasano, R., Buonavoglia, A., Pappagallo, F., et al. (2020). Actors on the scene: Immune cells in the myeloma niche. *Frontiers in Oncology, 10*, 599098. https://doi.org/10.3389/fonc.2020.599098.

Lewis, B. P., Burge, C. B., & Bartel, D. P. (2005). Conserved seed pairing, often flanked by adenosines, indicates that thousands of human genes are microRNA targets. *Cell, 120*, 15–20. https://doi.org/10.1016/j.cell.2004.12.035.

Li, Q., Ahuja, N., Burger, P. C., & Issa, J. P. (1999). Methylation and silencing of the Thrombospondin-1 promoter in human cancer. *Oncogene, 18*, 3284–3289. https://doi.org/10.1038/sj.onc.1202663.

Liu, L.-Z., Li, C., Chen, Q., Jing, Y., Carpenter, R., Jiang, Y., et al. (2011). MiR-21 induced angiogenesis through AKT and ERK activation and HIF-1α expression. *PLoS One, 6*, e19139. https://doi.org/10.1371/journal.pone.0019139.

Lyu, G., Guan, Y., Zhang, C., Zong, L., Sun, L., Huang, X., et al. (2018). TGF-β signaling alters H4K20me3 status via miR-29 and contributes to cellular senescence and cardiac aging. *Nature Communications, 9*, 2560. https://doi.org/10.1038/s41467-018-04994-z.

Mahtouk, K., Moreaux, J., Hose, D., Rème, T., Meissner, T., Jourdan, M., et al. (2010). Growth factors in multiple myeloma: A comprehensive analysis of their expression in tumor cells and bone marrow environment using Affymetrix microarrays. *BMC Cancer, 10*, 198. https://doi.org/10.1186/1471-2407-10-198.

Malcovati, L., Hellström-Lindberg, E., Bowen, D., Adès, L., Cermak, J., Del Cañizo, C., et al. (2013). Diagnosis and treatment of primary myelodysplastic syndromes in adults: Recommendations from the European LeukemiaNet. *Blood, 122*, 2943–2964. https://doi.org/10.1182/blood-2013-03-492884.

Marks, P. A. (2007). Discovery and development of SAHA as an anticancer agent. *Oncogene, 26*, 1351–1356. https://doi.org/10.1038/sj.onc.1210204.

Matsumura, S., Oue, N., Mitani, Y., Kitadai, Y., & Yasui, W. (2007). DNA demethylation of vascular endothelial growth factor-C is associated with gene expression and its possible involvement of lymphangiogenesis in gastric cancer. *International Journal of Cancer, 120*, 1689–1695. https://doi.org/10.1002/ijc.22433.

Miki, K., Shimizu, E., Yano, S., Tani, K., & Sone, S. (2001). Demethylation by 5-aza-2′-deoxycytidine (5-azadC) of p16INK4A gene results in downregulation of vascular endothelial growth factor expression in human lung cancer cell lines. *Oncology Research, 12*, 335–342. https://doi.org/10.3727/096504001108747783.

Mills, K. I., Kohlmann, A., Williams, P. M., Wieczorek, L., Liu, W., Li, R., et al. (2009). Microarray-based classifiers and prognosis models identify subgroups with distinct clinical outcomes and high risk of AML transformation of myelodysplastic syndrome. *Blood, 114*, 1063–1072. https://doi.org/10.1182/blood-2008-10-187203.

Miyamoto, N., Yamamoto, H., Taniguchi, H., Miyamoto, C., Oki, M., Adachi, Y., et al. (2007). Differential expression of angiogenesis-related genes in human gastric cancers with and those without high-frequency microsatellite instability. *Cancer Letters, 254*, 42–53. https://doi.org/10.1016/j.canlet.2007.02.004.

Muramatsu, F., Kidoya, H., Naito, H., Sakimoto, S., & Takakura, N. (2013). microRNA-125b inhibits tube formation of blood vessels through translational suppression of VE-cadherin. *Oncogene*, *32*, 414–421. https://doi.org/10.1038/onc.2012.68.

Nakagawa, S., Okabe, H., Ouchi, M., Tokunaga, R., Umezaki, N., Higashi, T., et al. (2018). Enhancer of zeste homolog 2 (EZH2) regulates tumor angiogenesis and predicts recurrence and prognosis of intrahepatic cholangiocarcinoma. *HPB: The Official Journal of the International Hepato Pancreato Biliary Association*, *20*, 939–948. https://doi.org/10.1016/j.hpb.2018.03.018.

Olsen, E. A., Kim, Y. H., Kuzel, T. M., Pacheco, T. R., Foss, F. M., Parker, S., et al. (2007). Phase IIb multicenter trial of vorinostat in patients with persistent, progressive, or treatment refractory cutaneous T-cell lymphoma. *Journal of Clinical Oncology*, *25*, 3109–3115. https://doi.org/10.1200/JCO.2006.10.2434.

Oronsky, B., Scicinski, J., Cabrales, P., & Minchinton, A. (2016). RRx-001, an epigenetic-based radio- and chemosensitizer, has vascular normalizing effects on SCCVII and U87 tumors. *Clinical Epigenetics*, *8*, 53. https://doi.org/10.1186/s13148-016-0220-7.

Oue, N., Matsumura, S., Nakayama, H., Kitadai, Y., Taniyama, K., Matsusaki, K., et al. (2003). Reduced expression of the TSP1 gene and its association with promoter hypermethylation in gastric carcinoma. *Oncology*, *64*, 423–429. https://doi.org/10.1159/000070302.

Piekarz, R. L., Frye, R., Turner, M., Wright, J. J., Allen, S. L., Kirschbaum, M. H., et al. (2009). Phase II multi-institutional trial of the histone deacetylase inhibitor romidepsin as monotherapy for patients with cutaneous T-cell lymphoma. *Journal of Clinical Oncology*, *27*, 5410–5417. https://doi.org/10.1200/JCO.2008.21.6150.

Pohlmann, P., DiLeone, L. P., Cancella, A. I., Caldas, A. P. F., Dal Lago, L., Campos, O., et al. (2002). Phase II trial of cisplatin plus decitabine, a new DNA hypomethylating agent, in patients with advanced squamous cell carcinoma of the cervix. *American Journal of Clinical Oncology*, *25*, 496–501. https://doi.org/10.1097/00000421-200210000-00015.

Qian, D. Z., Wang, X., Kachhap, S. K., Kato, Y., Wei, Y., Zhang, L., et al. (2004). The histone deacetylase inhibitor NVP-LAQ824 inhibits angiogenesis and has a greater antitumor effect in combination with the vascular endothelial growth factor receptor tyrosine kinase inhibitor PTK787/ZK222584. *Cancer Research*, *64*, 6626–6634. https://doi.org/10.1158/0008-5472.CAN-04-0540.

Qu, W., Wang, Y., Wu, Q., Liu, J., & Hao, D. (2015). Emodin inhibits HMGB1-induced tumor angiogenesis in human osteosarcoma by regulating SIRT1. *International Journal of Clinical and Experimental Medicine*, *8*, 15054–15064.

Raimondi, L., Amodio, N., Di Martino, M. T., Altomare, E., Leotta, M., Caracciolo, D., et al. (2014). Targeting of multiple myeloma-related angiogenesis by miR-199a-5p mimics: In vitro and in vivo anti-tumor activity. *Oncotarget*, *5*, 3039–3054. https://doi.org/10.18632/oncotarget.1747.

Raj, K., & Mufti, G. J. (2006). Azacytidine (Vidaza(R)) in the treatment of myelodysplastic syndromes. *Therapeutics and Clinical Risk Management*, *2*, 377–388. https://doi.org/10.2147/tcrm.2006.2.4.377.

Rao, X., Zhong, J., Zhang, S., Zhang, Y., Yu, Q., Yang, P., et al. (2011). Loss of methyl-CpG-binding domain protein 2 enhances endothelial angiogenesis and protects mice against hind-limb ischemic injury. *Circulation*, *123*, 2964–2974. https://doi.org/10.1161/CIRCULATIONAHA.110.966408.

Ribatti, D., Vacca, A., Nico, B., Presta, M., & Roncali, L. (2003). Angiogenesis: Basic and clinical aspects. *Italian Journal of Anatomy and Embryology*, *108*, 1–24.

Rigolin, G. M., Saccenti, E., Bassi, C., Lupini, L., Quaglia, F. M., Cavallari, M., et al. (2016). Extensive next-generation sequencing analysis in chronic lymphocytic leukemia at diagnosis: Clinical and biological correlations. *Journal of Hematology & Oncology*, *9*, 88. https://doi.org/10.1186/s13045-016-0320-z.

Roccaro, A. M., Sacco, A., Maiso, P., Azab, A. K., Tai, Y.-T., Reagan, M., et al. (2013). BM mesenchymal stromal cell-derived exosomes facilitate multiple myeloma progression. *The Journal of Clinical Investigation, 123*, 1542–1555. https://doi.org/10.1172/JCI66517.

Saltarella, I., Desantis, V., Melaccio, A., Solimando, A. G., Lamanuzzi, A., Ria, R., et al. (2020). Mechanisms of resistance to anti-CD38 daratumumab in multiple myeloma. *Cell, 9*, 167. https://doi.org/10.3390/cells9010167.

Saltarella, I., Lamanuzzi, A., Desantis, V., Di Marzo, L., Melaccio, A., Curci, P., et al. (2022). Myeloma cells regulate miRNA transfer from fibroblast-derived exosomes by expression of lncRNAs. *The Journal of Pathology, 256*, 402–413. https://doi.org/10.1002/path.5852.

Saunders, L. R., Sharma, A. D., Tawney, J., Nakagawa, M., Okita, K., Yamanaka, S., et al. (2010). miRNAs regulate SIRT1 expression during mouse embryonic stem cell differentiation and in adult mouse tissues. *Aging (Albany NY), 2*, 415–431. https://doi.org/10.18632/aging.100176.

Savage, K. J., Monti, S., Kutok, J. L., Cattoretti, G., Neuberg, D., De Leval, L., et al. (2003). The molecular signature of mediastinal large B-cell lymphoma differs from that of other diffuse large B-cell lymphomas and shares features with classical Hodgkin lymphoma. *Blood, 102*, 3871–3879. https://doi.org/10.1182/blood-2003-06-1841.

Shen, H., & Laird, P. W. (2013). Interplay between the cancer genome and epigenome. *Cell, 153*, 38–55. https://doi.org/10.1016/j.cell.2013.03.008.

Shi, H., Guo, J., Duff, D. J., Rahmatpanah, F., Chitima-Matsiga, R., Al-Kuhlani, M., et al. (2007). Discovery of novel epigenetic markers in non-Hodgkin's lymphoma. *Carcinogenesis, 28*, 60–70. https://doi.org/10.1093/carcin/bgl092.

Smith, Z. D., & Meissner, A. (2013). DNA methylation: Roles in mammalian development. *Nature Reviews. Genetics, 14*, 204–220. https://doi.org/10.1038/nrg3354.

Solimando, A. G., Annese, T., Tamma, R., Ingravallo, G., Maiorano, E., Vacca, A., et al. (2020). New insights into diffuse large B-cell lymphoma pathobiology. *Cancers (Basel), 12*, E1869. https://doi.org/10.3390/cancers12071869.

Solimando, A. G., Da Vià, M. C., Cicco, S., Leone, P., Di Lernia, G., Giannico, D., et al. (2019). High-risk multiple myeloma: Integrated clinical and omics approach dissects the neoplastic clone and the tumor microenvironment. *Journal of Clinical Medicine, 8*, 997. https://doi.org/10.3390/jcm8070997.

Solimando, A. G., Da Vià, M. C., Leone, P., Borrelli, P., Croci, G. A., Tabares, P., et al. (2020). Halting the vicious cycle within the multiple myeloma ecosystem: Blocking JAM-A on bone marrow endothelial cells restores angiogenic homeostasis and suppresses tumor progression. *Haematologica, 106*, 1943–1956. https://doi.org/10.3324/haematol.2019.239913.

Solimando, A. G., Da Vià, M. C., Leone, P., Borrelli, P., Croci, G. A., Tabares, P., et al. (2021). Halting the vicious cycle within the multiple myeloma ecosystem: Blocking JAM-A on bone marrow endothelial cells restores angiogenic homeostasis and suppresses tumor progression. *Haematologica, 106*, 1943–1956. https://doi.org/10.3324/haematol.2019.239913.

Solimando, A. G., Melaccio, A., Vacca, A., & Ria, R. (2022). The bone marrow niche landscape: A journey through aging, extrinsic and intrinsic stressors in the haemopoietic milieu. *Journal of Cancer Metastasis and Treatment*. https://doi.org/10.20517/2394-4722.2021.166.

Solimando, A. G., Ribatti, D., Vacca, A., & Einsele, H. (2016). Targeting B-cell non Hodgkin lymphoma: New and old tricks. *Leukemia Research, 42*, 93–104. https://doi.org/10.1016/j.leukres.2015.11.001.

Subramaniam, A. V., Yehya, A. H. S., Cheng, W. K., Wang, X., & Oon, C. E. (2019). Epigenetics: The master control of endothelial cell fate in cancer. *Life Sciences, 232*, 116652. https://doi.org/10.1016/j.lfs.2019.116652.

Sundaram, P., Hultine, S., Smith, L. M., Dews, M., Fox, J. L., Biyashev, D., et al. (2011). p53-responsive miR-194 inhibits thrombospondin-1 and promotes angiogenesis in colon cancers. *Cancer Research, 71*, 7490–7501. https://doi.org/10.1158/0008-5472.CAN-11-1124.

Szczyrba, J., Nolte, E., Hart, M., Döll, C., Wach, S., Taubert, H., et al. (2013). Identification of ZNF217, hnRNP-K, VEGF-A and IPO7 as targets for microRNAs that are down-regulated in prostate carcinoma. *International Journal of Cancer, 132*, 775–784. https://doi.org/10.1002/ijc.27731.

Theilgaard-Mönch, K., Boultwood, J., Ferrari, S., Giannopoulos, K., Hernandez-Rivas, J. M., Kohlmann, A., et al. (2011). Gene expression profiling in MDS and AML: Potential and future avenues. *Leukemia, 25*, 909–920. https://doi.org/10.1038/leu.2011.48.

Treon, S. P., Xu, L., & Hunter, Z. (2015). MYD88 mutations and response to Ibrutinib in Waldenström's Macroglobulinemia. *The New England Journal of Medicine, 373*, 584–586. https://doi.org/10.1056/NEJMc1506192.

Turunen, M. P., & Ylä-Herttuala, S. (2011). Epigenetic regulation of key vascular genes and growth factors. *Cardiovascular Research, 90*, 441–446. https://doi.org/10.1093/cvr/cvr109.

Waddington, C. H. (1959). Canalization of development and genetic assimilation of acquired characters. *Nature, 183*, 1654–1655. https://doi.org/10.1038/1831654a0.

Wang, Z.-F., Liao, F., Wu, H., & Dai, J. (2019). Glioma stem cells-derived exosomal miR-26a promotes angiogenesis of microvessel endothelial cells in glioma. *Journal of Experimental & Clinical Cancer Research, 38*, 201. https://doi.org/10.1186/s13046-019-1181-4.

Wang, Y., Sun, Z., & Szyf, M. (2017). S-adenosyl-methionine (SAM) alters the transcriptome and methylome and specifically blocks growth and invasiveness of liver cancer cells. *Oncotarget, 8*, 111866–111881. https://doi.org/10.18632/oncotarget.22942.

White, B. S., & DiPersio, J. F. (2014). Genomic tools in acute myeloid leukemia: From the bench to the bedside. *Cancer, 120*, 1134–1144. https://doi.org/10.1002/cncr.28552.

Wu, S. Y., Rupaimoole, R., Shen, F., Pradeep, S., Pecot, C. V., Ivan, C., et al. (2016). A miR-192-EGR1-HOXB9 regulatory network controls the angiogenic switch in cancer. *Nature Communications*, 7, 11169. https://doi.org/10.1038/ncomms11169.

Yan, B., Cheng, L., Jiang, Z., Chen, K., Zhou, C., Sun, L., et al. (2018). Resveratrol inhibits ROS-promoted activation and glycolysis of pancreatic stellate cells via suppression of miR-21. *Oxidative Medicine and Cellular Longevity, 2018*, 1346958. https://doi.org/10.1155/2018/1346958.

Yang, L.-T., Li, X.-X., Qiu, S.-Q., Zeng, L., Li, L.-J., Feng, B.-S., et al. (2016). Micro RNA-19a suppresses thrombospondin-1 in CD35+ B cells in the intestine of mice with food allergy. *American Journal of Translational Research, 8*, 5503–5511.

Yang, Q.-W., Liu, S., Tian, Y., Salwen, H. R., Chlenski, A., Weinstein, J., et al. (2003). Methylation-associated silencing of the thrombospondin-1 gene in human neuroblastoma. *Cancer Research, 63*, 6299–6310.

Yue, X., Wang, P., Xu, J., Zhu, Y., Sun, G., Pang, Q., et al. (2012). MicroRNA-205 functions as a tumor suppressor in human glioblastoma cells by targeting VEGF-A. *Oncology Reports, 27*, 1200–1206. https://doi.org/10.3892/or.2011.1588.

Zent, C. S., & Burack, W. R. (2014). Mutations in chronic lymphocytic leukemia and how they affect therapy choice: Focus on NOTCH1, SF3B1, and TP53. *Hematology. American Society of Hematology. Education Program, 2014*, 119–124. https://doi.org/10.1182/asheducation-2014.1.119.

Zhang, X., Dong, J., He, Y., Zhao, M., Liu, Z., Wang, N., et al. (2017). miR-218 inhibited tumor angiogenesis by targeting ROBO1 in gastric cancer. *Gene, 615*, 42–49. https://doi.org/10.1016/j.gene.2017.03.022.

Zhang, Y., Liu, J., Lin, J., Zhou, L., Song, Y., Wei, B., et al. (2016). The transcription factor GATA1 and the histone methyltransferase SET7 interact to promote VEGF-mediated angiogenesis and tumor growth and predict clinical outcome of breast cancer. *Oncotarget*, 7, 9859–9875. https://doi.org/10.18632/oncotarget.7126.

Zhang, Y., Wang, X., Xu, B., Wang, B., Wang, Z., Liang, Y., et al. (2013). Epigenetic silencing of miR-126 contributes to tumor invasion and angiogenesis in colorectal cancer. *Oncology Reports*, *30*, 1976–1984. https://doi.org/10.3892/or.2013.2633.

FSC
www.fsc.org
MIX
Papier | Fördert
gute Waldnutzung
FSC® C083411

Kolibri 360 GmbH
Ferdinand-Jühlke-Straße 7
99095 Erfurt, Deutschland
produktsicherheit@kolibri360.de